U.S. Department of Commerce
Technology Administration
National Institute of Standards and Technology

Office of Applied Economics
Building and Fire Research Laboratory
Gaithersburg, Maryland 20899-8603

Cost Analysis of Inadequate Interoperability in the U.S. Capital Facilities Industry

Michael P. Gallaher and Alan C. O'Connor
RTI International
Health, Social, and Economics Research
Research Triangle Park, NC 27709

John L. Dettbarn, Jr. and Linda T. Gilday
Logistics Management Institute
McLean, VA 22102

Prepared For:
> Robert E. Chapman
> Office of Applied Economics
> Building and Fire Research Laboratory
> National Institute of Standards and Technology
> Gaithersburg, MD 20899-8603

Sponsored By:
> National Institute of Standards and Technology
> Advanced Technology Program
> Information Technology and Electronics Office

Under Contract SB1341-02-C-0066

August 2004

I0487751

U.S. DEPARTMENT OF COMMERCE
Donald L. Evans, Secretary

TECHNOLOGY ADMINISTRATION
Philip J. Bond, Under Secretary for Technology

NATIONAL INSTITUTE OF STANDARDS AND TECHNOLOGY
Arden L. Bement, Jr., Director

Foreword

Information technologies have transformed many aspects of our daily lives and revolutionized industries in both the manufacturing and service sectors. Within the construction industry, the changes have so far been less radical. However, the use of information technologies offers the potential for revolutionary change in the effectiveness with which construction-related activities are executed and the value they add to construction industry stakeholders. Recent exponential growth in computer, network, and wireless capabilities, coupled with more powerful software applications, has made it possible to apply information technologies in all phases of the building/facility life cycle, creating the potential for streamlining historically fragmented operations.

Computer, automobile, and aircraft manufacturers have taken the lead in improving the integration of design and manufacturing, harnessing automation technology, and using electronic standards to replace paper for many types of documents. Unfortunately, the construction industry has not yet used information technologies as effectively to integrate its design, construction, and operational processes. There is still widespread use of paper as a medium to capture and exchange information and data among project participants.

Inadequate interoperability increases the cost burden of construction industry stakeholders and results in missed opportunities that could create significant benefits for the construction industry and the public at large. The lack of quantitative measures of the annual cost burden imposed by inadequate interoperability, however, has hampered efforts to promote the use of integration and automation technologies in the construction industry.

To address this need, the Building and Fire Research Laboratory and the Advanced Technology Program at the National Institute of Standards and Technology (NIST) have commissioned a study to identify and estimate the efficiency losses in the U.S. capital facilities industry resulting from inadequate interoperability among computer-aided design, engineering, and software systems. Although the focus of the study is on capital facilities—commercial/institutional buildings and industrial facilities—it will benefit key stakeholders throughout the construction industry.

This report, prepared for NIST by RTI International and the Logistic Management Institute, estimates the cost of inadequate interoperability in the U.S. capital facilities industry to be $15.8 billion per year. The intended audiences are owners and operators of capital facilities; design, construction, operation and maintenance, and other providers of professional services in the capital facilities industry; and public- and private-sector research organizations engaged in developing interoperability solutions.

The material contained in this report will promote an increased awareness of interoperability-related issues—both challenges and opportunities—in the capital facilities industry. The report addresses the cost burden issue by presenting both quantitative and qualitative findings and identifying significant opportunities for improvement. The report also analyzes the barriers to improved interoperability in the capital facilities industry and recommends actions for NIST and others to address these barriers.

Robert E. Chapman
Office of Applied Economics
Building and Fire Research Laboratory
National Institute of Standards and Technology
Gaithersburg, MD 20899-8603

Abstract

Interoperability problems in the capital facilities industry stem from the highly fragmented nature of the industry, the industry's continued paper-based business practices, a lack of standardization, and inconsistent technology adoption among stakeholders. The objective of this study is to identify and estimate the efficiency losses in the U.S. capital facilities industry resulting from inadequate interoperability. This study includes design, engineering, facilities management and business processes software systems, and redundant paper records management across all facility life-cycle phases. Based on interviews and survey responses, $15.8 billion in annual interoperability costs were quantified for the capital facilities industry in 2002. Of these costs, two-thirds are borne by owners and operators, which incur most of these costs during ongoing facility operation and maintenance (O&M). In addition to the costs quantified, respondents indicated that there are additional significant inefficiency and lost opportunity costs associated with interoperability problems that were beyond the scope of our analysis. Thus, the $15.8 billion cost estimate developed in this study is likely to be a conservative figure.

KEYWORDS

Building economics, interoperability costs, life-cycle cost analysis, capital facilities, electronic building design.

DISCLAIMER

Certain trade names and company products are mentioned in the text to adequately specify the technical procedures and equipment used. In no case does such identification imply recommendation or endorsement by NIST, RTI International, or LMI, nor does it imply that the products are necessarily the best available for the purpose.

Contents

Figures

Tables

Acronyms

3-D	three-dimensional
A&E or A&Es	Architects and Engineers (stakeholder group) or architecture and engineering [firms]
AEC	Architecture, Engineering, and Construction
AEX	Automating Equipment Information Exchanges with XML
ASPs	Application Service Providers
BFRL	Building and Fire Research Laboratory (NIST)
BLIS	Building Lifecycle Interoperable Software [group]
BLS	Bureau of Labor Statistics
BOMA	Building Owners and Managers Association
CAD	computer-aided design
CAE	computer-aided engineering
CALS	Continuous Acquisition and Life-Cycle Support
CAM	computer-aided manufacturing
CAx	computer-aided [design, engineering, etc.] system
CBECS	Commercial Buildings Energy Consumption Survey
CFMA	Construction Financial Management Association
CII	Construction Industry Institute
CIS/2	CIMSteel Integration Standards/Version 2
COMSPEC	Automated Specifications
CONCOM	Construction Communications
COs	change orders
CSI	Construction Specifications Institute
CSRF	Construction Sciences Research Foundation, Inc.
DoD	U.S. Department of Defense
EA	engineering and architectural [firm]
EDM	electronic document management

EIA	Energy Information Administration
EPCs	engineer/procure/construct
ERP	enterprise resource planning
FTEs	full-time equivalent workers
GC or GCs	General Contractors (stakeholder group)
GDP	gross domestic product
GIS	Geographic Information System
GSA	U.S. General Services Administration
HR	human resources
HTML	Hypertext Markup Language
HVAC	heating, ventilation, and air conditioning
IAI	International Alliance for Interoperability
IFCs	Industry Foundation Classes
IGES	Initial Graphics Exchange Specification
ISO	International Organization for Standardization
IT	information technology
LANs	Local Area Networks
MECS	Manufacturing Energy Consumption Survey
NAICS	North American Industry Classification System
NRC	National Research Council
O&M	operations and maintenance
OES	Occupational Employment Statistics
OO or OOs	Owners and Operators (stakeholder group)
R&D	research and development
REITs	real estate investment trusts
RFI	request for information
RFID	radio frequency identification
SF or SFs	Specialty Fabricators and Suppliers (stakeholder group)
SGML	Standard Generalized Markup Language
SRI	The Stanford Research Institute
STEP	STandard for the Exchange of Product Model Data
WWW	World Wide Web
XML	Extensible Markup Language

Executive Summary

The objective of this study is to identify and estimate the efficiency losses in the U.S. capital facilities industry resulting from inadequate interoperability. This study includes design, engineering, facilities management, and business processes software systems and redundant paper records management across all facility life-cycle phases. The capital facilities industry is changing with the introduction of information technology tools that have the potential to revolutionize the industry and streamline historically fragmented operations. These tools include computer-aided drafting technologies, 3-D modeling technologies, and a host of Internet- and standards-based design and project-collaboration technologies.

> In 2002, the value of capital facilities set in place in the United States was $374 billion. Even small improvements in efficiency potentially represent significant economic benefits.

Interoperability is defined as the ability to manage and communicate electronic product and project data between collaborating firms' and within individual companies' design, construction, maintenance, and business process systems. Interoperability problems in the capital facilities industry stem from the highly fragmented nature of the industry and are further compounded by the large number of small companies that have not adopted advanced information technologies.

Many manufacturing sectors, such as the automotive and aerospace industries, are in the process of harnessing emerging technologies to increase the efficiency of their design and manufacturing processes. Similar efficiency improvements that leverage automation and improved information flow have also been a topic of discussion within the capital facilities industry.

---

[1] U.S. Census Bureau. 2004b. "Annual Value of Construction Set In Place." As released on April 1, 2004 at http://www.census.gov/const/C30/Total.pdf.

> This study quantified approximately $15.8 billion in annual interoperability costs in the U.S. capital facilities industry, representing between 1 and 2 percent of industry revenue.[2] However, this is likely to be only a portion of the total cost of inadequate interoperability.

To inform the study, RTI International (RTI) and Logistics Management Incorporated (LMI) conducted a series of focus groups, telephone interviews and on-site interviews, and recruited organizations to participate in an Internet survey to develop interoperability cost estimates. Seventy organizations contributed data, anecdotes, and insights that informed the methodology and created the data set that led to this report's interoperability cost estimates. Many organizations had multiple individuals participate; thus, the number of individuals providing information for the study far exceeded the number of organizations.

Based on interviews and survey responses, $15.8 billion in annual interoperability costs were quantified for the capital facilities industry in 2002. Of these costs, two-thirds are borne by owners and operators, which incur these costs predominantly during ongoing facility operation and maintenance (O&M). In addition to the costs quantified, respondents indicated that there are additional significant inefficiency and lost opportunity costs associated with interoperability problems that were beyond the scope of our analysis. Thus, the $15.8 billion cost estimate developed in this study is likely to be a conservative figure.

E.1 DEFINING INTEROPERABILITY COSTS

The cost of inadequate interoperability is quantified by comparing current business activities and costs with a hypothetical counterfactual scenario in which electronic data exchange, management, and access are fluid and seamless. This implies that information need only be entered into electronic systems only once, and it is then available to all stakeholders instantaneously through information technology networks on an as-needed basis.

The concept of fluid and seamless data management encompasses all process data directly related to the construction and facility management process, including initial designs, procurement information, as-builts, and engineering specifications for O&M. The difference between the current and counterfactual scenarios represents the total economic loss associated with inadequate interoperability.

---

[2] Construction revenue includes the value of construction work and other business receipts for work done by establishments during the year (see Table 6-2).

Interoperability relates to both the exchange and management of electronic information, where individuals and systems would be able to identify and access information seamlessly, as well as comprehend and integrate information across multiple systems.

Examples of inefficiencies resulting from inadequate interoperability include manual reentry of data, duplication of business functions, and the continued reliance on paper-based information management systems. For the context of this analysis, three general cost categories were used to characterize inadequate interoperability: avoidance costs, mitigation costs, and delay costs.

- Avoidance costs are related to the *ex-ante* activities stakeholders undertake to prevent or minimize the impact of technical interoperability problems before they occur.

- Mitigation costs stem from *ex-post* activities responding to interoperability problems. Most mitigation costs result from electronic or paper files that had to be reentered manually into multiple systems and from searching paper archives. Mitigation costs in this analysis may also stem from redundant construction activities, including scrapped materials costs.

- Delay costs arise from interoperability problems that delay the completion of a project or the length of time a facility is not in normal operation.

E.2 METHODOLOGY FOR ESTIMATING COSTS

Our estimation approach focused on identifying and quantifying the interoperability *efficiency loss* associated with construction-related activities. During our interviews we also investigated *opportunity losses* associated with interoperability problems, but these costs are not included in the quantitative analysis because of their highly speculative nature. Our analysis approach is to determine costs that can be reliably documented, realizing that it is likely to result in an underestimate of total interoperability costs.

The economic methodology employed facilitated the quantification of annualized costs for 2002 that reflect interoperability problems throughout the construction life cycle. Costs are categorized with respect to where they are incurred in the capital facility supply chain. In turn, some portion of these costs is passed along in the form of higher prices. This study does not attempt to assess the impact on profits or consumer surplus; this would require market analysis to estimate changes in prices and quantities. However, in the long run economic theory suggests that all cost increases are eventually passed on to the final consumers of products and services.

For the purpose of this study, capital facilities industry encompasses the design, construction, and maintenance of large commercial, institutional, and industrial buildings, facilities, and plants.

Construction projects and facility operations are segmented into four life-cycle phases. In addition, interoperability problems affect an array of stakeholders and encompass a large number of activities. Thus our estimation procedure is built on a three-dimensional (3-D) framework (see Figure ES-1):

- *Facility Life Cycle*: design and engineering, construction, O&M, and decommissioning;

- *Stakeholder Groups*: aggregated to architects and engineers, general contractors, specialty fabricators and suppliers, and owners and operators;[3] and

- *Activities Categories*: efficiency losses from activities incurring avoidance, mitigation, and delay costs.

Average cost estimates per square foot were then calculated by life-cycle phase, stakeholder group, and cost category. These per-unit impacts were then weighted by construction activity or capital facility stock to develop national impact estimates for the capital facility industry. Total new construction activity for 2002 was estimated to be approximately 1.1 billion square feet (106 million square meters). The total square footage set in place was estimated to be nearly 37 billion (3.6 billion square meters). These estimates were developed using source data from the Energy Information Administration (EIA, 1997; EIA, 1998; EIA, 2001b; EIA, 2002).

E.3 PRIMARY DATA COLLECTION

One hundred and five interviews representing 70 organizations contributed to the estimation of inadequate interoperability costs. Invitations to participate in this study were distributed by a variety of means. Announcements were made at industry conferences and meetings. In addition, several trade associations and industry consortia issued notifications to their members via their Web sites, newsletters, periodicals, and word of mouth. Several organizations also participated in preliminary interviews to help define the scope of this project; these organizations continued their participation through the entirety of the effort.

---

[3] To make the scope of the project manageable, tenants were not included in the study. Tenants bear productive losses associated with downtime or suboptimal building performance. Because these direct costs are not included in the impact estimates, the total cost on inadequate interoperability is likely to be greater than the costs quantified in this study.

Figure ES-1. 3-D Representation of Estimation Approach of Inadequate Interoperability Costs

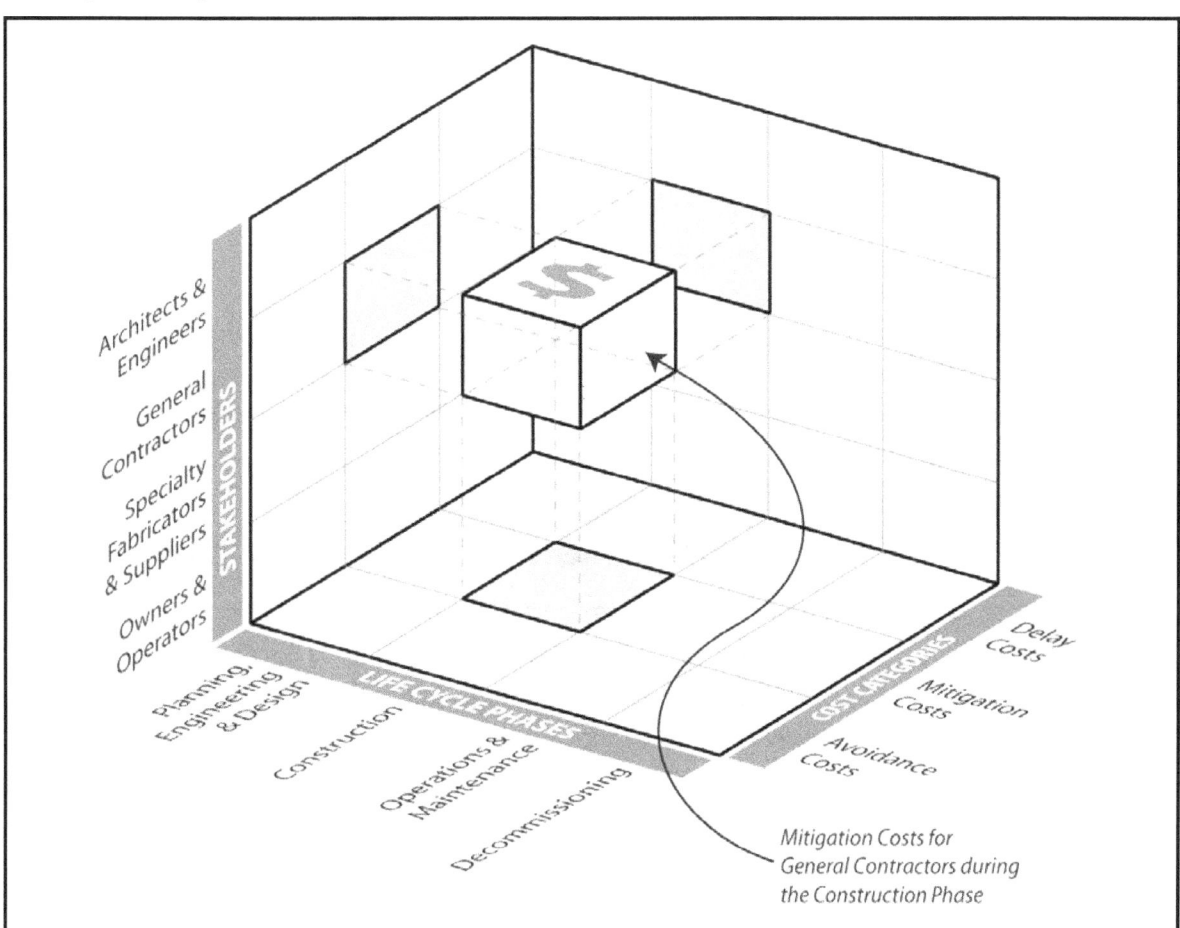

As shown in Table ES-1, owners and operators were the best represented stakeholder group with 28 organizations participating. Architects and engineers were represented by 19 organizations. Fourteen general contractors and specialty fabricators and suppliers organizations participated in the study. In addition, software vendors and research consortia contributed information concerning software applications, trends, and usage and on-going research and development efforts aiming to improve interoperability.

Table ES-1. Project Participants by Stakeholder Group

Stakeholder Group	Number of Interviewees	Number of Organizations
Architects and Engineers	21	19
General Contractors	11	9
Specialty Fabricators and Suppliers	5	5
Owners and Operators	53	28
Software Vendors	5	2
Research Consortia	10	7
Total	**105**	**70**

E.4 INTEROPERABILITY COSTS ESTIMATES

Based on interviews and survey responses, $15.8 billion in interoperability costs were quantified for the U.S. capital facilities supply chain in 2002 (see Table ES-2).[4] The majority of the estimated costs were borne by owners and operators; the O&M phase has higher costs associated with it than other life-cycle phases as information management and accessibility hurdles hamper efficient facilities operation. Owners and operators bore approximately $10.6 billion, or about two-thirds of the total estimated costs in 2002. Architects and engineers had the lowest interoperability costs at $1.2 billion. General contractors and specialty fabricators and suppliers bore the balance of costs at $1.8 billion and $2.2 billion, respectively.[5]

As shown in Table ES-3, most costs fall under the categories of mitigation and avoidance costs. Owners and operators primarily incur mitigation costs, and general contractors and special fabricators and suppliers primarily incur avoidance costs. Quantified delay costs are primarily associated with owners and operators. However, all stakeholder groups indicated that seamless exchange of electronic data would shorten design and construction time, even though many could not always quantify the impact.

---

[4] The term "quantify" is used when discussing the results to emphasize that data could not be collected to estimate all interoperability costs. Thus, the cost impacts presented in this section represent a subset of the total interoperability costs.

[5] We were unable to accurately assess the costs of the decommissioning phase because of limited survey responses. However, qualitative discussions indicate that the lack of reliable as-built and maintained information leads to a significant amount of resurveying and exploratory activities.

Table ES-2. Costs of Inadequate Interoperability by Stakeholder Group, by Life-Cycle Phase (in $Millions)

Stakeholder Group	Planning, Design, and Engineering, Phase	Construction Phase	Operations and Maintenance Phase	Total
Architects and Engineers	1,007.2	147.0	15.7	1,169.8
General Contractors	485.9	1,265.3	50.4	1,801.6
Specialty Fabricators and Suppliers	442.4	1,762.2	—	2,204.6
Owners and Operators	722.8	898.0	9,027.2	10,648.0
Total	**2,658.3**	**4,072.4**	**9,093.3**	**15,824.0**

Source: RTI estimates. Sums may not add to totals due to independent rounding.

Table ES-3. Costs of Inadequate Interoperability by Cost Category, by Stakeholder Group (in $Millions)

Cost Category	Avoidance Costs	Mitigation Costs	Delay Costs
Architects and Engineers	485.3	684.5	—
General Contractors	1,095.40	693.3	13.0
Specialty Fabricators and Suppliers	1,908.40	296.1	—
Owners and Operators	3,120.00	6,028.20	1,499.80
Total	**6,609.10**	**7,702.00**	**1,512.80**

Source: RTI estimates. Sums may not add to totals due to independent rounding.

E.5 TRENDS AND DRIVERS

Interviews with participants also included prospective discussions that focused on barriers to information management, communication, and exchange within the capital facilities supply chain and the opportunities that exist to eliminate these inefficiencies. Owners and operators in particular were able to illustrate the challenges of information exchange and management due to their involvement in each phase of the facility life cycle. In summary, they view their interoperability costs during the O&M phase as a failure to manage activities upstream in the design and construction process. Poor communication and maintenance of as-built data, communications failures, inadequate standardization, and inadequate oversight during each life-cycle phase culminate in downstream costs. This can be seen in the quantification of substantial costs related to inefficient business process management and losses in productivity for O&M staff.

However, owners and operators were not the only ones to express such frustrations regarding the costs they bear. During interviews with the three other stakeholder groups many of the same issues were discussed. They expressed the view that interoperability costs do not simply result from a failure to take advantage of emerging technologies, but rather, stem from a series of disconnects and thus a lack of incentives to improve interoperability, both within and among organizations, that contribute to redundant and inefficient activities.

1 Introduction to the Capital Facilities Industry

The capital facilities industry, a component of the entire U.S. construction industry, encompasses the design, construction, and maintenance of large commercial, institutional, and industrial buildings, facilities, and plants.[1] In 2002, the nation set in place $374 billion in new construction on capital facilities (U.S. Census Bureau, 2004b). The scope of this evaluation is limited to the capital facilities industry because the industry's large-scale projects and sophisticated data requirements make it particularly susceptible to interoperability problems. Therefore, it is hypothesized that the industry accrues the most significant share of the greater construction industry's interoperability costs.

In this study, the capital facilities industry covers construction-related activities and their associated supply chains throughout the life cycle of commercial, institutional, and industrial facilities. Thus, the industry includes such stakeholder groups as architects, engineers, general contractors, suppliers, and owners and operators. These stakeholders work in tandem to design, construct, operate, and decommission capital facilities.

The majority of this report is devoted to analyzing the issues and cost drivers that define the extent of interoperability issues the capital facilities industry faces. This chapter provides contextual information that formulates a framework for approaching the industry and that

---

[1] This discussion draws on two previous studies published by the National Institute for Standards and Technology (NIST) that define and detail the size and composition of the capital facilities industry. Chapman (2000) includes information on the industrial facilities portion of the capital facilities industry. Chapman (2001) includes information on the commercial buildings portion of the capital facilities industry

subsequently facilitates an understanding of the rationale that underlies the economic methodology for quantifying efficiency losses detailed in later chapters.

1.1 CAPITAL FACILITIES' SIZE AND SCOPE

A substantial portion of U.S. gross domestic product (GDP) is invested in capital facilities each year. Over $374 billion was invested in new facilities or facility renovations and additions in 2002 (U.S. Census Bureau, 2004b). Table 1-1 presents the annual value of construction set in place, according to the U.S. Census Bureau, for 1998 through 2002. The Census-defined facility categories depicted in this table best represent those facilities that are included within the scope of this analysis: nonresidential buildings and facilities. Capital facilities are broken out into three broad categories: commercial, institutional, and industrial facilities. This definition of capital facilities excludes transportation infrastructure such as bridges and roads to maintain a manageable scope for the project. However, it is apparent that these sectors also have significant needs for improved interoperability. The remainder of this section presents this analysis's definition of the commercial, institutional, and industrial facilities categories.

Table 1-1. Annual Value of Construction Put in Place, 1998–2002 ($Millions)

Type of Construction	1998	1999	2000	2001	2002
Office	42,226	47,582	55,605	52,260	37,578
Hotels	14,816	15,951	16,293	14,490	10,285
Religious	6,594	7,371	8,019	8,385	8,217
Other Commercial	53,598	56,915	60,381	60,760	55,879
Educational	44,699	50,724	57,018	62,626	68,995
Hospital and Institutional	17,716	17,600	19,219	19,153	22,366
Public Housing and Redevelopment	5,187	5,146	4,927	5,096	5,507
Industrial	41,494	33,564	32,959	31,077	18,486
Electric Light and Power	12,381	14,585	22,038	23,803	24,789
Public Utilities	39,212	44,066	50,994	53,849	54,370
Military	12,591	15,117	16,955	17,899	18,284
All Other Nonresidential	43,652	46,825	52,768	50,883	49,362
Total	**334,166**	**355,446**	**397,176**	**400,281**	**374,118**

Source: U.S. Census Bureau, 2004b.

Table 1-2 presents information on the number of buildings and total floor space for each facility category based on data gathered from the Energy Information Administration's (EIA) Commercial Building Energy Consumption Survey (CBECS) and the Manufacturing Energy Consumption Survey (MECS). Although the primary goal is to track energy consumption, these two surveys collect floor space and building population data, which was used in the analysis. The following discussions present information based on CBECS and MECS data on all commercial, institutional, and industrial facilities. However, our analysis focused on major facilities in each category, such as skyscrapers or hospitals.

Table 1-2. Total Square Footage for Existing Commercial, Institutional, and Industrial Facilities, 1998 and 1999

Principal Building Activity	All Buildings (thousands)	Total Floor Space (million square feet)	Total Floor Space (million square meters)
Commercial (1999)	**2,865**	**37,589**	**3,492.0**
Food sales	174	994	92.3
Food service	349	1,851	172.0
Lodging	153	4,521	420.0
Mercantile	667	10,398	966.0
Office	739	12,044	1,118.9
Public assembly	305	4,393	408.1
Service	478	3,388	314.7
Institutional (1999)	**833**	**16,142**	**1,499.6**
Education	327	8,651	803.7
Health care	127	2,918	271.1
Public order and safety	72	1,168	108.5
Religious worship	307	3,405	316.3
Industrial (1998)	**226**	**12,836**	**1,192.5**
Total	**3,924**	**66,567**	**6,184.5**

Note: Floor space statistics were originally reported in square feet. Units were converted to the metric system per NIST adoption of standard international units.

Source: EIA, 2001a; EIA, 2002.

1.1.1 Commercial Facilities

CBECS classifies buildings according to their primary business activity; buildings used for more than one activity are classified by the activity that uses the largest share of floor space for a particular building. The commercial sector includes office buildings and service businesses (e.g., retail and wholesale stores, hotels and motels, restaurants, and hospitals). Office buildings include general, professional, or administrative office space. Commercial facilities also include assembly buildings such as theaters, sports arenas, and meeting halls.

In1999 the United States had 739,000 office buildings representing just over 12 billion square feet (1.1 billion square meters) and 667,000 shopping areas representing nearly 10.4 billion square feet (1 billion square meters) of facilities. Overall, a total of nearly 37.6 billion square feet (3.5 billion square meters) were associated with the commercial buildings sector (EIA, 2002).

1.1.2 Institutional Facilities

Institutional buildings are defined as buildings used for the purpose of public services aimed at improving social welfare; this definition primarily includes large facilities dedicated to education, health care, and religious worship. This discussion presents data for institutional buildings separately from the EIA's usual joint commercial/institutional building classification in CBECS.

Educational facilities included buildings used for academic or technical classroom instruction, representing 327,000 buildings and 8.6 billion square feet (803.7 million square meters) in 1999. Campus buildings not used specifically for classroom instruction are classified elsewhere according to the principal activity for that building. Health care includes buildings used for the diagnosis and treatment of patients and accounted for 127,000 buildings and 2.9 billion square feet (271.1 million square meters). Religious worship includes buildings designed for religious gatherings and related activities (EIA, 2002). In 1999, the entire institutional category included 883,000 buildings totaling 16.1 billion square feet (1.5 billion square meters).

1.1.3 Industrial Facilities

The industrial sector consists of establishments that manufacture commodities as well as public utilities and large energy-producing establishments. Table 1-3 lists the major categories for industrial

Table 1-3. Industrial Enclosed Floor Space and Number of Establishment Buildings, 1998

NAICS Code[a]	Subsector and Industry	Establishments	Approximate Enclosed Floor Space of All Buildings Onsite (million square feet)	Approximate Enclosed Floor Space of All Buildings Onsite (million square meters)
311	Food	16,553	800	74.3
312	Beverage and tobacco products	1,547	205	19.0
313	Textile mills	2,935	363	33.7
314	Textile product mills	4,216	176	16.4
315	Apparel	12,566	258	24.0
316	Leather and allied products	995	39	3.6
321	Wood products	11,663	378	35.1
322	Paper	4,676	601	55.8
323	Printing and related support	25,782	417	38.7
324	Petroleum and coal products	1,756	88	8.2
325	Chemicals	8,962	1,237	114.9
326	Plastics and rubber products	11,944	855	79.4
327	Nonmetallic mineral products	11,333	435	40.4
331	Primary metals	3,830	600	55.7
332	Fabricated metal products	40,743	1,326	123.2
333	Machinery	19,577	1,031	95.8
334	Computer and electronic products	9,925	656	60.9
335	Electrical equipment, appliances, and components	4,526	1,350	125.4
336	Transportation equipment	8,380	1,119	104.0
337	Furniture and related products	11,274	488	45.3
339	Miscellaneous	13,630	414	38.5
	Total	**226,813**	**12,836**	**1,192.5**

[a]The North American Industry Classification System (NAICS) has replaced the Standard Industrial Classification (SIC) system.

Note: Floor-space statistics were originally reported in square feet. Units were converted to the metric system per NIST adoption of standard international units.

Source: EIA, 2001a.

manufacturing plants and floor space estimates from the 1998 MECS. Paper manufacturers accounted for 601 million square feet (55.8 million square meters). The plastics and rubber products industry had 855 million square feet (79.4 million square meters) and over 12,000 facilities. Chemical manufacturing had 1.2 billion square feet (114.9 million square meters) of industrial workspace (EIA, 2001b); this number was distributed across 11 chemical sectors, with the largest shares represented by plastic materials and resins and other basic organic chemicals.

While area is an adequate measure for most light manufacturing, it is not representative of the scale for heavy industrial and utility facilities, whose size is more accurately characterized by capacity rather than area. For example, according to the EIA (2001c), there were 153 operable petroleum refineries in the United States that represented only 88 million square feet (8.2 million square meters). Barrels per day is a more appropriate measure of refinery facility size, and total refinery capacity is almost 17 million barrels per day. Likewise, the preferred measure for electric utilities is megawatt capacity. On average, each plant is capable of producing over 2,500 megawatt hours of electricity per day (EIA, 2001a).

However, organizations are reticent to provide details on actual capacity data. In addition, reliable information on the capacity of heavy industrial facilities is unavailable. Therefore, this analysis made use of floor space data to draw comparisons across industrial facilities and generate national impacts.

1.2 CAPITAL FACILITIES STAKEHOLDERS

The breadth of the capital facilities industry includes a large array of stakeholders. The construction industry is fragmented and subject to many influences. Stakeholders include capital facility owners and operators, design and engineering firms, customers and end-users, constructors, suppliers and fabricators, technology vendors, governmental regulatory bodies, special interest groups, and governmental legislative bodies. Labor unions, trade and professional associations, research organizations and consortia, and even lobbyists play supporting roles.

To simplify the approach, this analysis focused on four stakeholder groups that directly accrue inadequate interoperability efficiency losses:

- Architects and Engineers (A&E), covering architects, general and specialty engineers, and facilities consultancies.

- General Contractors (GC), covering general contractors tasked with physical construction and project management.

- Specialty Fabricators and Suppliers (SF), covering specialty constructors and systems suppliers, including elevators, steel, and HVAC systems, for example.

- Owners and Operators (OO), covering the entities that own and/or operate the facilities.

In addition to these four primary groups, this analysis also investigated interoperability issues for consortia and research organizations, information technology vendors, and legal and insurance companies. These latter groups are also able to provide substantive input on how interoperability problems manifest themselves from a perspective different from those actively engaged in facilities management and construction.

To make the scope of the project manageable, tenants were not included in the study. Tenants bear productivity losses associated with downtime or suboptimal building performance. Because these direct costs are not included in the impact estimates, the total cost on inadequate interoperability is likely to be greater than the costs quantified in this study.

Each stakeholder group is affected in different ways by inadequate system interoperability. The following section provides an overview of the facility life cycle and the role stakeholders typically have in the design, construction, and operation of capital facilities.

1.2.1 Architects and Engineers

A&E firms design various capital facilities for private or public sector clients. These firms are often involved in multiple phases of the life cycle for a capital facility (BLS, 2003a). Architectural, design, and engineering firms work closely with the OOs, SFs, and GCs to ensure that specifications and objectives are met during construction.

There are three types of design firms: strictly architectural firms, architecture and engineering firms, and engineering and architectural firms. Architectural firms specialize strictly in the design of buildings (Gale Group, 2001a). These firms outsource engineering expertise. Architecture and engineering firms' primary competency is in architectural design, but they also employ structural engineers to contribute expertise to the design phase. Engineering and architecture firms focus primarily on engineering design services and employ a small number of architects (Gale Group, 2001a). This report refers to these three categories collectively as "A&E firms."

The design process has four stages. The first stage—design programming—allows the OO to decide the type of structure to build. The second phase is site selection and acquisition. At this stage, A&E firms consider various facts such as local tax rate, congestion, and topographical land features to decide where to build the selected structure. The third stage is conceptual design. Detailed models, both two- (2-D) and three-dimensional (3-D), are constructed to convey various design ideas, and to develop a hypothetical plan that can be used for cost estimating and to guide more detailed model building at a later stage. Finally, the architectural firm develops detailed documents in cooperation with engineers, ensuring that regional structural codes are met. These drawings require large amounts of highly detailed information that specifies the quality of materials and enables construction contractors to accurately bid on the project (Gale Group, 2001a).

Following the initiation of a construction project and, after approval of architectural and engineering designs, A&E firms spend most of their time coordinating information and any additional input from OOs and GCs (BLS, 2003a). A&E firms also spend time visiting the construction site to ensure that contractors are following design plans and that the project is running efficiently and within budget constraints.

The U.S. Census Bureau reports statistics for all architectural and engineering establishments, including those for residential construction (see Table 1-4). The values are reported at the five-digit NAICS level for 1997. The greater A&E industry comprised 73,128 establishments employing 876,750 workers and generating $105.2 billion in receipts in 1997. The number of establishments, employment size, and value of receipts pertaining to the capital facilities industry is a subset of the values reported in Table 1-4. The data in the table provide some measure of the size of the A&E stakeholder group defined in this study.

In 1997, a reported 146,702 paid employees worked for 20,602 architectural services firms (see Table 1-4). The Bureau of Labor Statistics (BLS) reported that employment in A&E firms decreased by 2000 (BLS, 2003b). Like the other stakeholder groups in the construction industry, the architectural services industry's success

Table 1-4. Architectural and Engineering Services, 1997, by NAICS Code

NAICS Code	Description	Establishments	Dollar Value of Business Done ($Millions)	Annual Payroll ($Millions)	Paid Employees
54131	Architectural services	20,602	16,988.3	6,468.5	146,702
54133	Engineering services	52,526	88,180.7	35,337.9	730,048
	Total	**73,128**	**105,169.0**	**41,806.4**	**876,750**

Source: U.S. Census Bureau, 2000b.

follows a standard business cycle. As capital availability is constrained by economic conditions, the number of new facility design projects also decreases.

Engineering services accounted for the largest proportion in terms of employment, number of establishments, and receipts (Table 1-4). However, these numbers include all types of engineering services firms from aerospace to environmental conservation. In reference to the construction industry, architectural firms act as a lead in designing structures and engineers are used as consultants to determine structural limits, feasibility of design, and process engineering (Gale Group, 2001a).

As Table 1-4 shows, this stakeholder group is composed of many firms. While some of the industry's larger firms compete nationally, most companies compete on a regional level (Tulacz, Rubin, and Armistead, 2003). Therefore, national comparisons and concentration measures are inadequate indicators of competition for the A&E stakeholder group.

A&E services are also highly fragmented, specializing in designing spaces for a wide range of sectors and industries. McGraw-Hill reported national market shares for "design firms" by facility type based on receipts from the top 500 firms (Tulacz, Rubin, and Armistead, 2003). McGraw-Hill's term "design firm" captured A&Es and combined architecture, engineering, and construction (AE&C) firms.

Table 1-5 reports the national market share by facility type for design firms based on receipts from the top 500 firms in 2002. Based on reported revenues, corporate building design accounted for 20 percent of the market for construction-related design services. Petroleum-related facilities accounted for slightly more than 12 percent, whereas manufacturing and industrial facility design work accounted for only 2 and 6 percent, respectively.

Table 1-5. Architect and Engineering Service Revenues, 2002, by Facility Type

Type of Work	Revenue ($Millions)	Percent of Total
Corporate buildings	10,240	20.4
Manufacturing	1,268	2.5
Industrial	3,072	6.1
Petroleum	6,192	12.4
Water	2,968	5.9
Sewer/waste	3,669	7.3
Transportation	9,849	19.7
Hazardous waste	5,060	10.1
Power	4,943	9.9
Telecommunications	926	1.9
Other	1,831	3.8
Total	**48,186**	**100.0**

Source: Tulacz, Rubin, and Armistead, 2003.

1.2.2 General Contractors

GCs are construction execution specialists and coordinate closely with A&E and OO firms. Normally, a single GC specializing in one type of construction acts as the project coordinator during the build phase. The GC is responsible for all construction activities; however most project work is frequently subcontracted to heavy industrial and/or specialty trade contractors (BLS, 2003a).

GCs coordinate the construction process in cooperation with the A&Es' design plan and local building codes. GCs often have expertise in a certain type of facility construction such as educational, healthcare, petroleum, and commercial facilities (Gale Group, 2001c). For large construction projects, management responsibilities are often segmented into the various stages of the construction process, such as site preparation (e.g., land clearing and sewage systems), building construction (e.g., foundation and erection of the structural framework), and building systems installation (e.g., ventilation, electrical, fire, and plumbing) (BLS, 2003c).

Table 1-6 summarizes the GC stakeholder category. In 1997, 44,709 establishments employed 671,238 workers, which generated more than $209 billion in receipts. Commercial and institutional building contractor firms represented over 80 percent of the establishments for

Table 1-6. General Contracting Services, 1997, by NAICS Code

NAICS Code	Description	Number of Establishments	Dollar Value of Business Done ($Millions)	Annual Payroll ($Millions)	Paid Employees
23331	Manufacturing & industrial building construction	7,279	34,038.4	5,129.0	143,065
23332	Commercial & institutional building construction	37,430	175,230.8	19,176.2	528,173
	Total	**44,709**	**209,269.2**	**24,305.1**	**671,238**

Source: U.S. Census Bureau, 2000a. Sums may not add to totals due to independent rounding.

nonresidential construction, and employed 528,173 workers (see Table 1-6). However, over the next 2 years, commercial and institutional building contractors suffered a downturn, which reached a low point in 1999 (Gale Group, 2001b).

McGraw-Hill reports market shares by facility type based on the top 400 contractors' revenues (Tulacz and Powers, 2003). In 2002, building construction had the highest revenue in the market, accounting for over 50 percent of the total market (see Table 1-7). Power plant construction accounted for nearly 10 percent, petroleum facilities for 8 percent, and industrial construction for over 5 percent of the market. These four categories are largest in terms of revenue.

1.2.3 Specialty Fabricators and Suppliers

Individual SFs specialize in one particular trade and often work as subcontractors on task orders from a GC. SFs perform narrowly defined tasks within the major construction process and repairs following the completion of construction (BLS, 2003a). Examples of special trade contractors include heating and air conditioning contractors (NAICS 23822), structural steel erection contractors (NAICS 23812), and building equipment installation contractors (NAICS 23829), which include elevator contractors, for example. The broad SF category, defined as NAICS 235, Special Trade Contractors, by the U.S. Census Bureau, employed over 3.4 million people and performed $340.9 billion in business in 1997 (Census, 2000a). These figures include counts for both residential and nonresidential activity. The remainder of Section 1.2.3 discusses three SF subsectors as examples of the size and modes of work for this broad stakeholder category.

Table 1-7. General Contractor Revenue, 2002, by Facility Type

Type of Work	Revenue ($Millions)	Percent of Total
Building	98,336	50.6
Manufacturing	6,204	3.2
Industrial	10,114	5.2
Petroleum	15,872	8.2
Water	3,038	1.6
Sewer/waste	3,353	1.7
Transportation	25,849	13.3
Hazardous waste	6,279	3.2
Power	18,843	9.7
Telecommunications	2,706	1.4
Other	3,796	2.0
Total	**194,390**	**100.0**

Source: Tulacz and Powers, 2003.

Heating and air-conditioning contractors install, service, and repair climate-control systems in capital facilities (BLS, 2003e). Following equipment installation, additional infrastructure such as fuel and water supply lines, air ducts and vents, pumps, and other supporting equipment must also be installed. Heating and air-conditioning contractors work directly with the GCs during the construction phase. However, due to servicing requirements, heating and air-conditioning contractors also work closely with OOs over the life of a facility. SFs are often involved in more than one phase of the capital facility life-cycle and, therefore, coordinate and communicate with GCs, A&Es, and OOs.

The heating and air-conditioning contractors (NAICS 2351) group, which also includes plumbing contractors, accounted for over 20 percent of all establishments and 23 percent of employment in special trade contracting (see Table 1-8). The Bureau of Labor Statistics (BLS, 2003e) predicted that, through the year 2010, employment will rise faster for heating and air-conditioning installers than the average rate for other special trade contractors. While employment may suffer due to a slowing of new construction projects, servicing of existing systems and repair work should remain stable over time.

Table 1-8. Specialty Fabricators and Suppliers, 1997, by NAICS Code

NAICS Code	Description Special Trade Contractors	Establishments	Dollar Value of Business Done ($Millions)	Annual Payroll ($Millions)	Paid Employees
2351	Plumbing, heating, and air-conditioning contractors	84,876	88,427.4	25,720.2	788,930
2353	Electrical contractors	61,414	64,915.1	21,680.0	641,984
23591	Structural steel erection contractors	4,238	8,152.7	2,387.1	72,301
23592	Glass and glazing contractors	4,713	4,045.5	1,051.6	35,823
23594	Wrecking and demolition contractors	1,541	2,304.0	592.2	18,820
23595	Building equipment and other machinery installation contractors	4,488	9,342.9	3,148.0	75,501
	Total	**161,270**	**177,187.7**	**54,579.0**	**1,633,359**

Source: U.S. Census Bureau, 2000a. Sums may not add to totals due to independent rounding.

Structural steel erection services prepare the site by building cranes and steel frames used during the erection process to move materials around the construction site. Following site preparation, steel erection workers build the steel structural skeleton of the building. Steel beams arrive on the construction site in numbered sections, which are then lifted into position by a crane and attached to the existing structural skeleton (BLS, 2003f).

In 1997, the structural steel erection contractor subsector reported employment of 72,301 workers (see Table 1-8). This subsector represented 14 percent of employment in the other special trade contractor (NAICS 2359) segment. BLS predicts that employment in the special trade sector is expected to match the average increase for all SFs through 2010 (BLS, 2003f).

In 1997, building equipment and other machinery installation contractors, as defined by the Census, consisted of 4,488 establishments (see Table 1-8). The subsector employed 75,501 workers and performed over $9.3 billion in business. This subsector consists of several types of mechanical system installers, such as elevator installers.

Elevator installers install, maintain, and repair elevator systems. Installation requires familiarity with blueprints to determine the equipment necessary. Installation includes welding the rails to the existing building structure inside the elevator shaft; assembling the car's platform, walls, and doors; and installing rollers along the side of the car. These contractors also install outer doors at the elevator entrances at each floor in the facility (BLS, 2003d). Like heating and air-conditioning systems, elevator systems require continuing maintenance for the life of the equipment. Continuing maintenance requires elevator installation contractors to deal with information from GCs during construction of a facility and OOs over the lifetime of the facility.

The Bureau of Labor Statistics (BLS, 2003d) predicted that employment in this sector will grow at the average rate through 2010; but, as with all specialty trade contractors, employment growth depends on the rate of capital investment in real estate.

1.2.4 Owners and Operators

Unlike the previous stakeholders, which fall into well-defined industry categories, OOs are ubiquitous across all industry NAICS codes. Thus, the data presented in Table 1-9 are for illustrative purposes only. Any corporation or institution that owns, maintains, and/or operates a capital facility is considered an OO. This includes organizations as diverse as corporations, real estate management companies, the General Service Administration (GSA), and the Department of Defense, for example. This disparity presents some difficulty when discussing OOs at an aggregate level.

Census statistics present information on nonresidential real estate property managers (NAICS 531312), which are examples of nontenant OOs (see Table 1-9). Firms classified by these NAICS codes own and operate capital facilities, renting to various commercial, industrial, and institutional clients. In 1997, 53,525 establishments were involved in rental, leasing, and property management. These establishments employed 299,990 workers and generated almost $51.8 billion dollars in revenue. The 50 largest nonresidential property managers (NAICS 531312) accounted for over 27 percent of this sector's revenues. Real estate investment trusts (REITs) and the U.S. GSA are examples of OOs for the private and public sectors, respectively. In the late 1990s, REITs became a popular industry for facility management. REITs contract the design and construction of new facilities and also specialize in the acquisition of existing facilities (Gale Group, 2001d). In 2000, REITs owned 10 percent of all commercial and industrial facilities in the United States. Industrial facilities represented one-third of those owned, while retail accounted for one-fifth of the industry holdings (Gale Group, 2001d).

Table 1-9. Selected Owners and Operators, 1997, by NAICS Code

NAICS Code	Description	Establishments	Dollar Value of Business Done ($Millions)	Annual Payroll ($Millions)	Paid Employees
53112	Lessors of nonresidential buildings (except miniwarehouses)	31,497	38,105.1	3,828.4	145,317
53119	Lessors of other real estate property	12,017	5,539.3	685.6	37,623
531312	Nonresidential property managers	10,011	8,146.2	3,738.8	117,050
	Total	**53,525**	**51,790.6**	**8,252.8**	**299,990**

Source: U.S. Census Bureau, 2000c. Sums may not add to totals due to independent rounding.

The GSA manages over 1,700 government-owned facilities, accounting for over 55 percent of the federal government building inventory. GSA hires design and general contracting firms to build and maintain federal buildings such as court houses, office buildings, national laboratories, and data processing centers (GSA, 2003).

1.2.5   Fragmentation among Stakeholders

The stakeholder groups listed above represent well over 700,000 individual firms (U.S. Census Bureau, 2000a, b, c). Each stakeholder must be able to effectively communicate information and specifications to other stakeholders during the construction process. The market for services between and within each of the stakeholders is fragmented due to a large number of establishments, regional competition, and lack of incentives for coordination. In addition, there is frequently a lack of industry guidance and agreement on best practices and facility delivery strategy.

Regionalism characterizes the construction industry. Given that 8 out of 10 construction establishments have fewer than 10 employees (BLS, 2003a), competition is localized within a single region rather than nationally. Building codes, worker compensation, and facility type vary across regions, making national competition extremely costly for smaller firms (Tulacz, Rubin, and Armistead, 2003).

Lack of incentives for coordination also exists, which has contributed to fragmentation of the industry. OOs are facing increasing pressure internally to lower the costs associated with new and additional facility construction. In a recent study by McGraw-Hill, 83 percent of facility OOs interviewed cited poor project planning as a critical issue in cost of new facility construction (Tulacz and Rubin, 2002). Productivity is another key issue in curtailing costs of construction. OOs believe that the development and implementation of "better tools," such as information management software, and improved communication between A&Es and GCs are ways to improve construction productivity. Progress has been made to organize the technology development consortia led by some of the major OOs. However, many of these productivity enhancement suggestions have gone unrecognized by GCs (Tulacz and Rubin, 2002).

1.3 FACILITY LIFE-CYCLE PHASES

Capital facilities pass through a number of stages or phases: planning and design, construction, and commissioning; operations and use (to include maintenance and renewal/revitalization actions); and then decommissioning and disposal (Cleland, 1999; Hudson, Haas, and Uddin, 1997; NRC, 1998).[2] Therefore, the facility life cycle is segmented into four broad life-cycle phases. In general, these phases are applicable to all facility types. Any key differences are typically found in the level of regulatory oversight for the facility.

Figure 1-1 presents a diagram of the various phases of the facility life cycle; each of these phases is described in more detail in the following sections. Figure 1-2 is paired with Table 1-10, which presents the expected design life of selected types of facilities and infrastructure for comparison. However, some industrial owners do not base their investment decisions on such a long design life because the building's usefulness (product's predicted sales life) is far shorter. Thus, the investment planning life cycle may be shorter than 50 to 60 years in many cases.

As shown in Figure 1-1, the first two phases may take 2 to 5 years in a life cycle that may last a total of 45 to 50 years for a commercial building. This becomes significant because typically 30 to 40 percent of the total life-cycle costs for a facility occur in these first two phases and 60 to 70 percent in the third phase (in constant dollars). In other words, the operations and maintenance (O&M) costs tend to be the dominant costs of ownership for facilities and infrastructure, yet these costs are difficult

---

[2] Cotts (1998) uses a slightly different taxonomy to describe the life cycle: planning, acquisition, O&M, and disposition. The acquisition phase encompasses the concept, design, and construction activities addressed in the NRC, Hudson, and Cleland discussions.

Figure 1-1. Facility Life-Cycle Phases

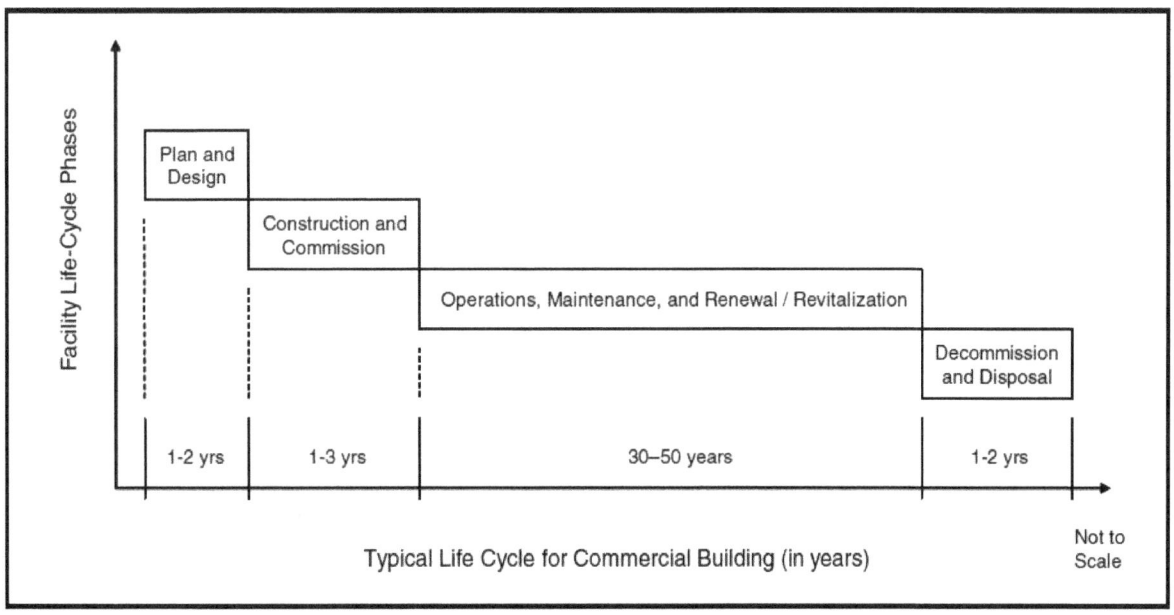

Source: LMI.

to factor into buy/sell decisions made over the life of the facility assets. Facility assets, unlike ships or airplanes, typically change hands one or more times during the life of the property. Owners typically focus on the design and construction costs of the facility and deal with O&M costs later in the budget cycle (Cotts, 1998; Cleland, 1999; DSMC, 2001; NRC, 1990; NRC, 1998; Sullivan, Wicks, and Luxhoj, 2003).

Figure 1-2 presents the cost impact of changing a design at various stages in the facility life cycle. A classic example of this impact occurs when the O&M aspects of a type of heating and ventilation equipment are not considered in the design phase. The designer may have designed the room housing the heating and ventilation equipment to meet existing space constraints and specified equipment that physically fits in the space allocated. When installed, the facility operators may find that, although the equipment is in the room, it cannot be properly maintained due to limited clearances between the equipment and walls of the room, thereby requiring a physical relocation of one or more of the walls. This type of problem is very expensive to rectify after construction is complete; resolving this conflict prior to construction is more efficient and involves less disruption. Better interoperability may alleviate such occurrences.

Figure 1-2. Cost Impact of Changing Facility Design at Differing Stages in the Facility Life Cycle

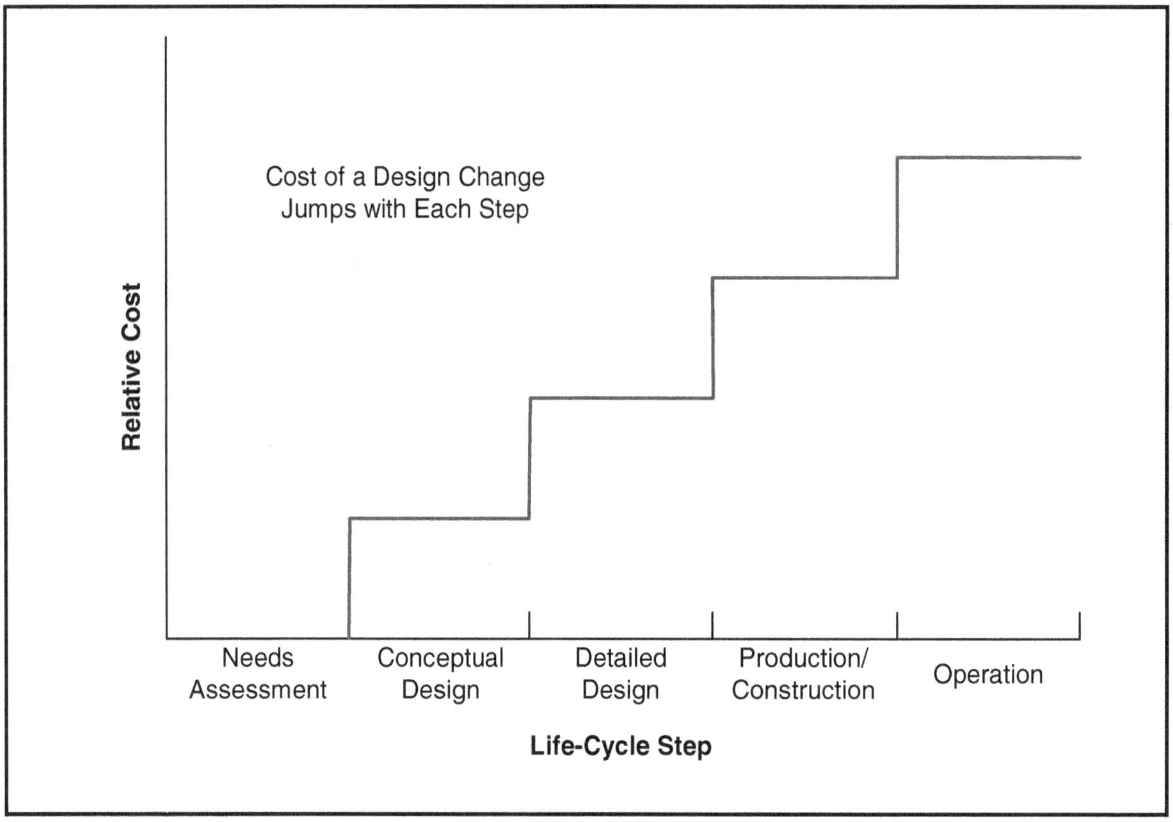

Source: LMI.

Table 1-10. Expected Design Life by Facility Type

Facility or Infrastructure Element	Expected Design Life (in years)
Commercial buildings	30 to 50 years
Industrial buildings	50 to 60 years
Utility systems	75 to 100 years

Sources: Cotts, 1998; Hudson, Haas, and Uddin, 1997; NRC, 1998.

Figure 1-3 presents the top-level business processes found in typical commercial and industrial facility construction projects. A discussion of each life-cycle phase follows. A more detailed description of the typical business process is presented in Appendix A.

Figure 1-3. Typical Facility Life Cycle for Commercial Building

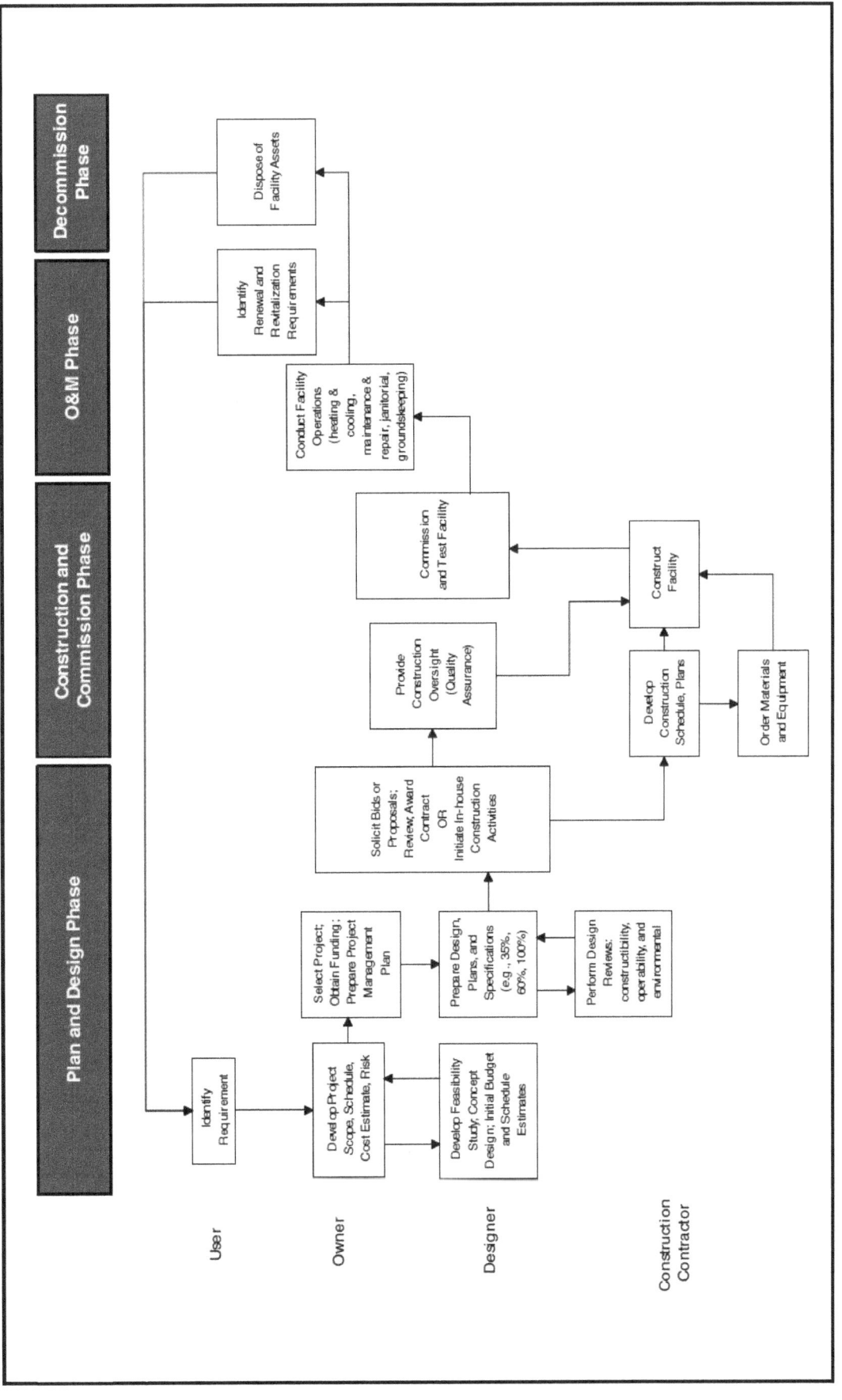

Source: LMI.

1.3.1 Planning, Engineering, and Design Phase

The planning, engineering, and design phase starts or initiates a construction project. Depending on the type of project to be undertaken, this phase commences many years prior to the opening of a new facility. Stakeholders are engaged in a suite of activities during this initial phase. A&Es, SFs, GCs, and OOs may all be involved because many of the decisions made at this stage significantly affect the following phases.

Several assessments must be undertaken before commencing construction. Stakeholders perform needs assessments to establish the need for, or identify, an investment opportunity for new construction. If the project is deemed viable, several activities are set into play, including

- a master plan for the project, including initial budgets;

- an evaluation of real estate options, including zoning and permitting;

- an environmental impact study to evaluate energy and resource use, toxic by-products, indoor air quality, and waste products, among others; and

- selection of consultants, A&Es, SFs, and GCs.

Stakeholders confer to develop detailed schedules and plans for completion of the facility. Subphases include schematic design, design development, detailed design engineering, construction documents, and permit and/or agency approvals. Permits for things such as soil evacuation, grading, drilling, building and/or equipment drainage, ceiling penetrations, asbestos work, work in confined spaces, hot work, hazardous work (explosives), lead work, radiation work, and roof access are sought and acquired prior to the start of construction. Building equipment and outfitting items are specified and ordered. As bids and quotes are received, they are reviewed by the owner. Occasionally, redesigns and new equipment specifications are needed to develop new designs and specifications that will reduce the budget. Work is performed within the parameters set in the feasibility study.

1.3.2 Construction Phase

Construction processes and activities include both new construction activities and those related to additions and alterations. New construction activities include the original building of structures, essential service facilities, and the initial installation of integral equipment, such as elevators and plumbing, heating, and air-conditioning supplies and equipment. Additions and alterations include construction work that adds to the value or useful life of an existing building or structure, or that adapts a building or structure to a new or different use. Also included are major replacements of building systems (e.g., installation of a new roof or heating system).

During this phase, the facility is built and transferred to the facility operators. The construction project is typically implemented through bidding, negotiations, and contract award. After a contract award is made, the GC develops detailed construction schedules; develops safety, health, and environmental plans; and aligns subcontractors to complete the work. Shop drawings are approved by the owner and construction materials (such as building equipment; lighting; heating, ventilation, and air conditioning [HVAC]; and control systems) are approved and then purchased and installed according to the plans. As construction proceeds, the owner's representative provides quality assurance oversight to ensure that the general contractor is satisfactorily meeting the requirements of the contract.

1.3.3 Operations and Maintenance Phase

During this phase of the life cycle, the owner operates and maintains the facility. Following testing, regulatory compliance, and confirmation of the project implementation and completion, the facility is commissioned and transferred to the owner. Furniture and outfitting items can be delivered and installed and, finally, the building can be occupied.

In the context of this study, facility operations included the activities required to provide necessary building services to the facility occupants, such as heating and cooling; building maintenance (preventive and corrective) and repair; space and move management; health, safety, and environmental management; and janitorial, grounds-keeping, pest control, and snow removal services.

The National Research Council (NRC, 1998) stated that the deteriorating condition of public sector facilities "is attributable, in part, to the failure to recognize the total costs of facilities ownership." As a facility ages during the many decades it is in service, periodic renewal or revitalization activities are needed, in addition to the facility operations described above. As Figure 1-4 depicts, a building's performance will decline because of its age, the use it receives, or functional adaptation to new uses, but its performance will decline at an optimized rate with proper maintenance. Without appropriate maintenance, or with the owner's

Figure 1-4. Maintenance Effect on Facility Performance

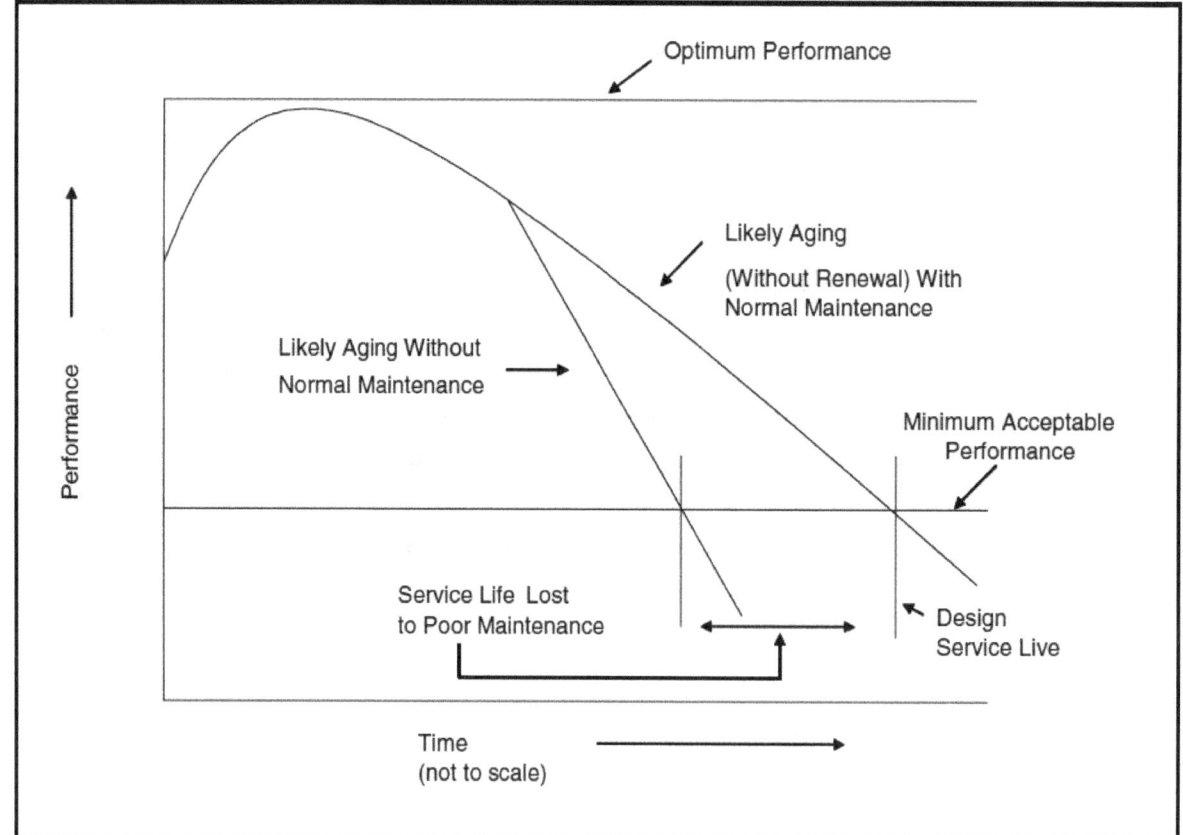

Source: NRC, 1998.

decision to defer required maintenance, the building's usefulness will decrease at an accelerated rate.

The total cost of ownership is the total of all expenditures an owner will make over a building's service lifetime. Failure to recognize these costs and to provide adequate maintenance, repair, and renewal results in a shorter service life, more rapid deterioration, higher operating costs, and possible mission degradation over the life cycle of a building. With available data on facility subsystems, an estimate can be made regarding maintenance, repair, and renewal requirements during the remaining asset lifetime. Managing this data is of critical importance to effectively provide optimum services to the facility owner and users.

The systems employed during this phase of the facility life cycle should interoperate with the planning and design systems to provide the most efficient data and information. Planning, design, construction, operation and renovations, and demolition decisions are made throughout a facility's life cycle and are based primarily on economic performance.

Owners and other stakeholders in the facility life cycle tend to make decisions based on the range of information available to them. Thus, inaccurate or poorly defined information impedes their ability to make sound economic decisions.

1.3.4 Decommissioning Phase

This phase of the life cycle occurs when facility use is terminated. It is characterized by transfer of equipment (if required) to new facilities, environmental clean-up, and disposal. The stakeholders in this phase typically include the owner and, when environmental restoration and clean-up are required, the federal and state regulators, local community, and others involved in the process. Decision options at this point include selling the facility or infrastructure asset, demolishing it, or abandoning it. Once a decision is made, a project is typically initiated to execute the alternative selected.

2 Evolution of Information Exchange in the Capital Facilities Industry

This chapter provides an historical overview of information exchange within the capital facilities industry. To evaluate historical and current means for information sharing, this chapter reviews the development of electronic information systems and the rate at which the industry adopted them. Many systems have been made available by information technology vendors, but stakeholders' adoption of them has been moderate. The traditional paper-based means of communication remain entrenched in the industry. Though industry stakeholders acted through several consortia to enhance the viability of using electronic systems and information exchange, to date these efforts have met with limited success.

2.1 THE INTRODUCTION OF ELECTRONIC SYSTEMS

Paper has been and still is the most common medium for storing and transferring information in the capital facilities industry. The introduction of computer use has done little to create the "paperless office." As discussed in Chapter 1, the capital facilities industry is complex and fragmented, subject to many actors and influences, with changing project ownership over each life-cycle phase. In fact, many decision makers in the owners' organizations may not fully appreciate the interrelated nature of the business processes that support the facility life cycle. It is from this foundation that system (or data) interoperability issues occur. A fragmented business process and organizational structure will tend to create fragmented and inefficient business and management systems.

The following discussions present a summary of electronic information exchange among the facility life-cycle phases. It is important to note that many of the software application types mentioned during one phase are applicable for other phases as well. For example, owners and operators employ computer-aided design, engineering, and manufacturing systems both during the planning, engineering, and design phase and during the operations and maintenance phase. To simplify the discussion, these systems are presented once.

2.1.1 Electronic Systems in the Planning, Engineering, and Design Phase

The first of the four life-cycle phases—planning, engineering, and design—has the highest volume of software tools and use of electronic systems relative to the amount of work performed. In the past three decades, there has been a trend to replace paper-based correspondence with electronic mail and also to introduce spreadsheet software for use in the initial planning stage to support budget preparation and cost control. In addition, technologies have been adopted to make business support functions more efficient.

Computer-Aided Design, Engineering, and Manufacturing

In the early 1970s, design professionals used computer-based specification development programs that operated on mainframe computers. With the introduction of smaller personal computers and the use of magnetic and optical media for information distribution, the programs were adapted to the hardware and software most commonly used by the industry for project development. The specification development programs were designed to manipulate, and be compatible with, construction information, cost databases, and product information libraries. This was a large step toward the modernization of construction industry communications. It allowed the computer to support design efforts and specify what should be built and how.

By the early 1980s, some design professionals and engineers prepared and made decisions on facilities using computer-aided drafting and design. When first adopted, computer-aided design (CAD) packages were used to replace tasks normally done by hand on paper (typically a drawing). But these systems were initially not a cost-effective investment because they were expensive, difficult to learn and use, and did not significantly improve productivity. In addition, early CAD systems were run using large mainframes and dedicated hardware systems.

Over time, however, stakeholder data requirements and innovation in software packages converged to make the use of CAD and, later, computer-aided engineering (CAE) and manufacturing (CAM) more efficient and economically viable. These software systems became less expensive, easier to use, and had more sophisticated applications. For instance, instead of only the 2–D views envisioned 20 years ago, software now exists that allows graphical 3-D view of designs. It should be noted, however, that there is a fundamental difference between drafting systems that are not really design systems at all but merely capture the results of design in terms of 2-D views and design systems from which 2-D and 3-D views can be extracted. Drafting systems cannot natively detect clashes, missing components, incompatible connections, inconsistencies between drawings, physically impossible configurations, and many other errors that plague design. Modeling systems can, so this is why they have already replaced drafting systems in complex system projects.

The revolution in personal computing during the 1980s and 1990s permitted users to run these applications from their work stations using desktop computers. Although there are still issues to be addressed regarding data transfer between multiple systems and interoperability issues related to interaction with a client, CAD software is largely accepted by the various stakeholder groups. In addition to physical design software, there exists a significant market for add-ons such as CAD-viewers, data translators, and rendering software that generates more realistic interpretations of a facility long before the physical construction begins.

The Construction Financial Management Association (CFMA) cites 35 different CAD software applications used by the construction industry. Examples include ArchiCAD, AutoCAD, CADsoft Build, MicroStation, Paydirt, and NavisWorks (CFMA, 2002).

Internet-Based Information Sharing and Collaborative Tools

During the 1990s, the Internet became very popular and provided a new communications medium over which to view and exchange information. For instance, the use of the universal Internet data formats and access technologies is replacing various proprietary interfaces. Similarly, Intranets have become a widely used tool for employees of the same firms to work together on projects from their own computers, at any time, rather than having to hold in-person collaborative meetings. Internet-based technology is applied to facilitate information exchange and the sharing of resources among project teams because internal and external parties can communicate and share data more quickly and effectively. These collaborative tools have access verification and control features that allow privacy to be maintained.

Internet project portals are a common collaboration tool gaining popularity in the project management arena. The value added by project portals and their customizable, central repositories of information is provided by their relative ease of use and Internet connectivity. Through the use of a collaborative tool such as project portals, an organization can use dispersed teams (sometimes known as virtual teams) to enhance communication among team members at different geographic locations.

Extranets are secure and private networks that use the Internet protocol and public telecommunication to securely share business information with other suppliers, vendors, and others. They can be viewed as external parts of a company's intranet.

Standardized Information and Formats

The development efforts for standardized information formats began decades before the Web was popular, starting with low-level efforts like the standardization of ASCII and continuing with mid- and high-level efforts like IGES, STEP, IFC, and CIS/2, all of which were driven by the availability of multiple, incompatible information systems. The benefits of electronic data exchange are pushing the demand for better data exchange formats. To date, there are no clear and straightforward choices for neutral file formats, although several organizations and standards bodies are developing them. At present, it is usually necessary for project teams to discuss at the beginning of the project which formats to use, and for each party to set up whatever data translation facilities may be needed and to test the conversions before the project is fully under way. Even with formats that provide useful exchange infrastructures, preliminary testing is essential to ensure proper interoperability.

XML development started in 1996 and was derived from SGML (developed in the early 1980s) and HTML (developed in 1990). Originally designed to meet the challenges of large-scale electronic publishing, it now plays a role in the exchange of a wide variety of data. XML is now the accepted language for data communication over the Internet. XML uses tags to communicate to a computer how to create and define elements within a data set and interpret the contents of electronic documents transferred (Blackman, 2001). The designers of XML created a set of guidelines or conventions for designing text formats to structure data. XML makes it easy for computers to generate data, read data, and ensure that the data structure is unambiguous. XML can be used to store any kind of structured information and to enclose or encapsulate information to pass it between different computing systems that would otherwise be unable to communicate (O'Reilly & Associates, Inc., 2003) and is license free.

An example of XML applied to the capital facilities industry is "aecXML," which is an XML-based language under development to represent information in the Architecture, Engineering and Construction (AEC) industry. The aecXML initiative, which originated at Bentley Systems, is now managed by the International Alliance for Interoperability (IAI). aecXML seeks to establish common schema definitions, using well-defined business cases, for AEC data via the standard XML formatting language (IAI-NA, 2003). aecXML is intended to support specific business-to-business transactions over the Internet. Such transactions may be associated with the transfer of resources such as project documents, materials, parts, and contact information. aecXML has the potential to enable greater efficiency for activities such as proposals, design, estimating, scheduling, and construction.

Segments of the capital facilities industry have been working to improve data exchange capabilities for the past 20 years. The following are included among other efforts:

- *Interim Graphics Exchange Specification.* In 1980, the Interim Graphics Exchange Specification (IGES) Organization was formed. It was the first effort that recognized the need to exchange product definition data rather than merely CAD data.

IGES allows different CAD/CAM systems to interchange product-definition data.

- *Standard for the Exchange of Product Model Data.* In the mid-1980s the manufacturing sector created a need for STandard for the Exchange of Product Model Data (STEP). It was the first effort that recognized the need to standardize product data representations before expressing them in a standard exchange syntax and format via application protocols. STEP, as a part of the ISO body of standards,[8] is a worldwide effort to develop a mechanism for exchanging and sharing engineering data. STEP works toward neutral industrial data definitions, representation, and language that supports life-cycle functions. The use of a common exchange format helps reduce translation costs and improve quality throughout the use of the data. STEP enables product data sharing between software applications throughout a product life cycle, different organizations involved in a product life cycle, and physically dispersed sites within an organization.

- *Industry Foundation Classes* (IFCs). IFCs, under development by IAI, are designed to provide a means of passing a complete, thorough, and accurate building data model from the computer application used by one participant to another with no loss of information. IFCs are data elements that represent the parts of buildings or elements of a process for a particular facility and contain the relevant information about those parts. Computer applications use IFCs to assemble a computer-readable model that constitutes an object-oriented database. This database may be shared among project participants and continue to grow as a project goes through design and construction and enters operation.[9] The European Council for Civil Engineers estimates that the use of IFCs can reduce the risk factors for facility management contractors by up to 20 percent for new buildings and up to 50 percent for older structures.[10]

The first set of IFCs was published in 1998 as Release 1.5. This was quickly replaced a year later by R2.0 and later by an extensible version R2x in 2000. The goal is to create a language that relates information on shape, component attributes, and the relationships between components. A recent *Engineering News Record* article explains how the IFC approach works:

> IFC compatible software typically incorporates third party tools that output platform-specific data in a universally understood EXPRESS-based format. A CAD file, normally saved in a native drawing format such as DWG or DGN, would also be output in an IFC format. An estimating or analysis program could then open the IFC file, recognize standard objects such as walls, window, and doors, and

---

[8]Designed by the ISO Technical Committee 184/SC4, it is called "ISO 10303: Industrial Automation Systems and Integration—Product Data Representation and Exchange."

[9]Adapted from http://www.iai-na.org (IAI-NA, 2003). For additional information on IFCs, view the IAI Web site or http://www.fiatech.org.

[10]See http://www.eccenet.org for additional information.

perform its own tasks on the same pool of information. (Roe and Reina, 2001)

In 1999, a group of software vendors interested in facilitating the adoption of IFC R2.0 formed the Building Lifecycle Interoperable Software (BLIS) group. Vendors participating in this project included Graphisoft, Timberline Software Corp., and Microsoft, in addition to several U.S. government-based OOs. The BLIS project demonstrated that project data "could be shared between various software products during design, energy analysis, quantity takeoff, and code checking" (Roe and Reina, 2001).

- *CIMSteel Integration Standards/Version 2* (CIS/2). CIS/2 is a protocol through which stand-alone programs, such as structural analysis, CAD, and detailing systems, can communicate with each other. By providing a neutral data format, CIS/2 allows data interchange between a wide variety of program types. CIS/2 is the logical product model and electronic data exchange format for structural steel project information. CIS/2 has been implemented in many steel design, analysis, engineering, fabrication, and construction applications to create a seamless and integrated flow of information among all parties of the steel supply chain involved in the construction of steel-framed structures.[11]

2.1.2 Electronic Systems in the Construction Phase

Prior to the more recent use of the Internet, some construction companies used document handling systems that allowed project documents, but not drawings, to be shared over closed computer systems using telephone dial-up systems. These products were not user-friendly, had few genuinely helpful functions, and could only be justified when used on major projects.

Even though many tools are available today, few construction projects are completed using a majority of electronic tools. From a practical standpoint, field personnel tend to take handwritten notes, read hard-copy plans, develop quantity take-offs manually, and develop drawings that reflect the actual constructed facility using a paper-based drawing set and red marker (known as "as-built red-line drawings"). GCs are reticent to convert to electronic systems. For instance, when used in the harsh weather environments found on most job sites, electronic tools are not as easy to transport, set up, see, or use collaboratively as is a 36- by 44-inch set of blueprints.

---

[11] For additional information on how CIS/2 has helped the steel industry, see http://www.aisc.org/cis2. For a discussion of how CIS/2 was developed, see NIST's discussion (NIST, 2002) and Georgia Tech's description (Eastman, 2001).

There is a recent trend to incorporate new technology, such as handhelds, cellular phones, modems, and other devices to increase the connectivity between offices and the field. This trend will continue as more firms, particularly OOs, understand and mandate the use of electronic devices to effect increases in productivity.

Design changes incur significant costs for the OO once fabrication of building components starts. Software applications helped to reduce design errors considerably in recent years and allow managers to thoroughly analyze the building for different purposes prior to construction start. For instance, prefabricated building components must fit together properly at the site, as some building systems require precise positioning of structural components (such as for piping connections). When prefabricated materials and components are manufactured with varying degrees of dimensional accuracy, problems arise. If a window opening in a prefabricated panel is made too large, it creates significant issues for the window installers. The adoption of CAD/CAM techniques helps overcome these types of dimensional variances.

Construction project managers and field personnel still maintain one or more paper versions of the following: specifications, design document (drawings), contract(s), or bills and invoices. Typical reasons given by project managers for using paper versions are that they

- view electronic work as being more costly;
- do not trust that information will not be lost by a "crashed" computer or power outage;
- maintain old habits ("that's the way we've always done it");
- routinely communicate with stakeholder partners who
  - do not have or cannot operate computers or the necessary software, or
  - do not have the same version or application that is required;
- must meet county, city, or state requirements of having paper formats and original certification stamps by the registered officials;
- know that it is difficult to carry a laptop to the construction trailer for use by more than one or two people at a time, impeding group discussions;
- predict that construction workers would not be able to effectively or accurately use electronic formats;
- perceive that using electronic media at job sites with employees not accustomed to using electronic media is inefficient;

- believe it is more official to have paper; and
- do not have incentive(s) to
  - maintain a seamless repository of information about the facility given the emphasis on reducing direct labor activity, or
  - work electronically.

Software applications that facilitate business processes during the construction phase are relatively new. Many of the same systems mentioned in the preceding discussion are used by stakeholders in the construction phase as well. In addition, stakeholders employ a wide range of software tools to automate many of the processes that were traditionally accomplished using paper-based systems.

Examples of business process software include estimating, job costing, accounting, payroll, project management, project scheduling, and collaboration software. This software is used primarily by stakeholders to gain or enhance efficiencies related to internal operations and/or interactions with other team members. Only recently have business process software applications become affordable to traditionally low-tech, small to mid-sized firms, allowing them to automate certain aspects of their business processes.

Thus, the extent and frequency of electronic systems used during the construction phase have varied. However, a variety of electronic tools are available that are designed to streamline operations: estimating, job costing/accounting/payroll, project management, project collaboration, and scheduling applications, among many. The Construction Financial Management Association (CFMA) reports in a recent survey that Excel is the most widely used software application for all business processes (CFMA, 2002). Other process-specific applications exist; the following discussion briefly touches on systems cited by the CFMA.

Examples of estimation software used in the construction industry include

- AccuBid,
- Bidmaster Plus,
- McCormick Estimating, and
- Precision Collection.

Job costing/accounting/payroll software examples include

- COINS,

- Forefront,
- Gold Collection, and
- Viewpoint.

Project management tools are predominately used by larger firms; most smaller firms use Excel to track and monitor activities. In general, larger firms are more likely to use task-specific software tools because of greater access to resources and larger overall project size or number. The prevalence of project-management software use specifically for design and as-built is less than that for other categories of software. For those firms that use project management software, examples include

- Primavera Enterprise,
- Primavera Expedition, and
- Prolog Manager.

Collaboration software has only recently begun to take hold in the construction industry, due in large part to mandates by the owners and operators in their bid requirements. Similar to project management tools, the construction industry for the most part does not use collaboration software. CFMA reported that only 25 percent of construction firms surveyed used collaboration software (CFMA, 2002). Examples include

- Buzzsaw,
- Constructware, and
- Meridian Project Talk.

Scheduling software is again similar to project management and collaboration software in that its use increases relative to firms' size. Examples of project scheduling software applications, for those firms that use this type of application, include

- Microsoft Project,
- Suretrak, and
- Primavera Enterprise.

2.1.3 Electronic Systems in the Operations and Maintenance and Decommissioning Phases

Once construction is complete, design and as-built drawing archives and project files containing equipment and outfitting information are passed to OOs. A facility manager typically "takes over" during this part of the life cycle and is forced to use what the GC transfers. One way to limit transfer problems is to have the OO stakeholders involved in the early planning for the design and construction phases of the work.

To manage the facility operations, many different commercial facility management software packages, called Computerized Maintenance Management Systems (CMMS), are available. Some are simple "out of the box" applications, and others are complex and powerful with extensive customization capabilities. These products are designed to minimize the time required to plan and implement facilities management decisions.

Single software packages enable facility managers to manage portfolios and incorporate scheduling applications (such as Microsoft® Project and Primavera) and CAD drawings for building plans, layouts, furniture, telecommunications, and electronic data. Systems also need to interface with human resources, financial, purchasing, project management, accounting, and asset management information systems. Systems may provide access to tabular data such as furniture and equipment inventories and hierarchical data such as organization charts and space planning schemes. These information types have typically required specialized tools to access and present the data in an appropriate format. While the tools required are still specialized, the details of using them are eliminated by the facilities management interface, which provides a consistent and nontechnical front to even the most technical of data.

Facilities management systems allow better control of repair and maintenance activities and allow areas of exceptional performance or under-performance to be clearly and quickly identified. These systems also allow detailed asset, performance, and service inventories that add value to the property owner's portfolio when ownership is transferred and the facility is eventually decommissioned.

2.1.4 Data Exchange Paradigm for Life-Cycle Phases

OOs typically hire the design and engineering team based on competitive bidding; team members consequently often vary from project to project, meaning that information systems are not standardized among team members. Instances where OOs stipulate in the bid documents that, if chosen as the successful GC, all parties must operate with the same hardware and version of a predetermined software package are becoming more frequent. This is typically done for very large projects but not for smaller ones.

A typical job may use software to produce design and engineering documents, but these documents are predominantly reviewed, approved, permitted, and used by field personnel in paper format. Once a design is complete and agreed to by the owner, paper versions govern and electronic copies are requested "for the file." A&Es submit documents in a similar manner (paper versions govern and electronic copies are requested for the file). Specifications may be developed with software that uses past project information. Information is input into spreadsheets and submitted electronically to the OO for approval. Drawings are submitted to trades for their bids in paper format. Paper versions are used predominantly in the field, and electronic ones may be used if the paper version cannot be found.

Extranets have increased AEC communication and added efficiencies but not to the level anticipated by many information systems users. There often seems to be some link in the process that does not work— many times the data integrity and reliability is a function of the data entry source. For instance, some organizations use disparate databases that are not interoperable. Critical data elements, such as cost, schedule, and quality performance, may need to be keyed in multiple times, thereby creating a multiple data entry problem. Often times, data entry is not performed in the field because of the press of time required to focus on other "more important" issues.

2.2 STANDARDIZATION EFFORTS TO INCREASE USE OF ELECTRONIC SYSTEMS AND PROCESS DATA

Process data is defined as the data describing a process required to support the complete life cycle of the product or service delivered, from requirements and concept design through decommissioning of the facility. As described in this chapter, the past three decades have seen a maturation of the data and information available to the owner and other stakeholders in the facility life-cycle process.

Other (nonfacility) industries have worked aggressively on process data issues. Continuous Acquisition and Life-Cycle Support (CALS) is an initiative designed to reengineer document data handling, within and across organizations, to improve data access and reduce life-cycle costs. CALS started with the U.S. Department of Defense (DoD) and has since expanded to commercial applications across Asia, Europe, and North America. The aim of CALS is to reduce the cost of supporting and maintaining information of all kinds, including technical documentation.

Because many organizations receive goods and services from a wide range of suppliers, contractors, and subcontractors, they regularly combine, republish, and "repurpose" massive quantities of technical information.

The solutions are not as well defined or contained for the capital facilities environment. There are numerous overlapping and incompatible project delivery and e-commerce systems related to engineering and construction, repeating electronically the inefficient and fragmented paper-based business processes of the capital facilities industry.

In addition, there has been minimal convergence on nomenclature, information exchange standards, and improved work packaging to support collaboration, automation, and integration for the design, delivery, and operation of capital facilities. Several concurrent research initiatives are underway to investigate process data creation and interoperability issues for the construction industry. The International Alliance for Interoperability (IAI) and FIATECH are examples of nonprofit organizations developing standardized data formats and lobbying for industry-wide adoption and implementation of these standards.

2.2.1 FIATECH

The capital facilities industry formalized the critical need for interoperability standards in the FIATECH Capital Facilities Technology Roadmap (2002) with recommendations for action on the Critical Capability: Integrated and Automated Procurement and Supply Networks. The Construction Industry Institute (CII) Project Team 180, eCommerce for Construction, reported at the August 2002 CII Annual Conference that leading adopters of e-commerce for capital facilities projects have not succeeded in exploiting this technology for the design and delivery of equipment.

The lack of interoperability standards is a primary barrier to improving efficiency in service delivery. In collaboration with FIATECH, NIST developed the plan for the "Automating Equipment Information Exchanges with XML" (AEX) Phase 1 Project. Twelve organizations, including NIST's Building and Fire Research Laboratory (BFRL), committed resources to the AEX Phase 1 project. The first workshops were held during the summer of 2002.

2.2.2   International Alliance for Interoperability

IAI has been working since 1995 to develop systems that attempt to label components and systems of building construction and performance, attaching a rigorous set of attributes, or characteristics and behaviors, to each type of object. The resulting definitions are called Industry Foundation Classes (IFC), as introduced in Section 2.1.1.

2.2.3   Other Efforts

The Construction Sciences Research Foundation, Inc. (CSRF), an independent, not-for-profit construction industry research organization, expends significant effort to unify, integrate, and facilitate communication between software programs used in facilities design and construction. CSRF was founded in 1967 by The Construction Specifications Institute (CSI) to implement the recommendations of The Stanford Research Institute (SRI) for development of Automated Specifications (COMSPEC) and Construction Communications (CONCOM).

The Building Lifecycle Interoperable Software (BLIS) project began in the late 1990s and was initiated to promote the implementation of the existing IAI IFCs in application software.[12] It was conceived as a way to initiate the next logical phase in the widespread adoption of an object data model standard for stakeholders. Today their goals are to

- deliver increasing levels of application interoperability through semantic model sharing (objects, properties and relationships— not line drawings) and implementation collaboration by subgroups working to support specific BLIS "views";

- "jump start" IFC support in shipping applications and IFC-based interoperability; and

- validate any proposed extensions to IFC through software implementation.[13]

Another effort is the UNIFORMAT II elemental classification for building specifications, cost estimating, and cost analysis. Developed by NIST BFRL and industry representatives, the UNIFORMAT II initiative provides a classification framework for consistent reference to major building functions, or elements, that perform a given function regardless of design specification, construction method, or materials. UNIFORMAT II is now ASTM (American Society for Testing and Materials) Standard E1557-02

---

[12]The continued development of the IFC is being pursued by the IAI.
[13]See http://www.blis-project.org/index2.html.

(ASTM, 2002). The authors identify the following benefits from the application of UNIFORMAT II (Bowen, Charette, and Marshall, 1992):

- Elemental cost estimates are faster and less costly to produce than detailed estimates.

- Data are entered in a consistent format, facilitating electronic tracking of the building and its components.

- Stakeholder coordination is improved because elements are linked using standardized naming conventions.

- Standardized formats for collecting and analyzing data save time and improve the quality of cost estimates.

- Building condition assessments are easier to perform.

- Performance specifications are more understandable because they are indicated using standardized terms.

3 Current State of Interoperability in the Capital Facilities Industry

This chapter uses the review of the capital facilities literature and preliminary interviews with industry stakeholders to identify the sources and impact of inadequate interoperability. The initial scoping interviews yielded insightful comments on the potential cost reductions that could be achieved if interoperability were improved in the capital facilities supply chain. In addition, the interviews identified preliminary solutions for improving interoperability in the future.

3.1 OVERVIEW OF INADEQUATE INTEROPERABILTY SOURCES

The construction industry, although it acknowledges the need to reduce project costs and time, continues to use dominant conventional processes that prevent potential improvements in interoperability from occurring. A review of the literature indicates that reductions in delivery time, on the order of 20 to 50 percent, are possible through the use of enabling technologies and improved communication between all stakeholders in the industry (Bayramoglu, 2001; Beck, 2001; Luiten and Tolman, 1997).

Chinowsky (2001) found that the construction industry is struggling to change its focus from short-term projects to long-term strategic planning with an emphasis on customers and the enterprise. Using the Fortune 500 as a model of strategic management benefits, he found that the construction industry performs long-term planning, competitive market analyses, and implementation planning 30 to 40 percent less often than the Fortune 500 companies.

There are several reasons the construction industry suffers from inefficiency in information management.[14] Many parties, each with expert knowledge in different disciplines, often operate in isolation and do not effectively communicate knowledge and information with teaming partners both internally and externally. Inefficiency sources include the following:

- Collaboration software is not integrated with other systems. Some companies use collaboration software. Though it is effective, it is usually a stand-alone application and is not integrated with other systems. Furthermore, many parties work together on only one project so there is little incentive to invest in long-term solutions. As each project is often unique, each project tends to have different partners, scope, workforce, teams, and location. It takes time to get teams performing well together.

- Life-cycle management processes are fragmented and are not integrated across the project life cycle. Yet coordination is essential. Many projects require a significant number of requests for information questions and drawings between owners, architects, contractors, and subcontractors, often as many as 300 to 500 on a typical project.

- There are inefficiencies and communication problems when stakeholders from all parts of the life cycle have either various versions of the same software or different software.

- CAD interoperability issues arise since data are difficult to manage between differing applications and platforms; for example, making legacy CAD drawings consistent and as-built data available to newer programs.

- A lack of data standards inhibits the transfer of data between different phases in the life cycle and their associated systems and applications. According to one stakeholder, a late 1980s study by a large oil company found potential savings of 11 to 14 percent of operations and maintenance funding if data were in a consistent structure.

- Internal business processes are fragmented and inhibit interfirm and intrafirm interoperability. Design, engineering, and operations systems are typically not integrated. In addition, it is rare to find legacy systems that communicate effectively with each other or with new systems. In some firms, an estimated 40 percent of engineering time is dedicated to locating and

---

[14]These issues are derived from the literature and from a series of in-depth interviews with key firms and practitioners in the construction industry.

validating information gathered from disparate systems. Data-centric solutions are needed to maintain the quality and reliability of facilities management data.

- Many firms use both automated and paper-based systems to manage data and information. In many cases, the hard-copy construction documents are used on the jobsite in lieu of electronic copies. Electronic versions, therefore, often do not reflect facilities' as-built specifications.

- Many smaller construction firms and some government agencies do not employ, or have limited use of, technology in managing their business processes and information.

3.2 IMPACT OF INADEQUATE INFORMATION EXCHANGE AND MANAGEMENT

Industry stakeholders participated in a preliminary round of informal discussions aimed at gaining an initial foothold on the breadth of the capital facility industry's interoperability issues. Stakeholders categorized the impact of inadequate information exchange into the following areas: lack of standards, ineffective communication processes, lack of communication between CAD and other information technology systems, limited re-use of project knowledge across the firm, decision analytics, lack of clear priorities, and paper management issues. Respondents reported that related efficiency losses can be stated in terms of cost, schedule, or manpower.

One respondent indicated that "our industry's inability to communicate effectively has created tremendous waste and inefficiency, estimated at up to 30 percent of the total cost of each building project. Today, advanced computer technologies based on object-oriented data provide us with an opportunity to create synergy among each discipline's language and make the industry information truly 'interoperable.'" Another respondent indicated that during the construction phase alone, approximately 10 percent could be saved as a result of improved project scheduling efficiencies. One A&E industry professional indicated that a 20 to 50 percent reduction range in delivery time would be achievable through adoption of new technologies and improved communication between all stakeholders in the industry.

One consortium created within the construction industry works from the premise that a 30 to 40 percent savings could be achieved, based on the results of improved interoperability in the manufacturing community. While few feel confident enough to state such quantitative industry-wide savings projections, others have estimated, with a higher degree of confidence, the impact of interoperability in terms of time, personnel, or cost savings comparisons.

A review of stakeholders' anecdotal comments constitutes the remainder of this chapter. To facilitate comparison with the economic methodology presented in Chapter 4, the following text is organized according to the avoidance, mitigation, and delay cost paradigm that best characterizes interoperability costs. These cost categories are fully explored in Chapter 4, but a brief introduction is helpful here. Avoidance costs are those that are incurred to prevent interoperability problems from occurring. Mitigation costs are those that are incurred to correct problems once they have occurred. Finally, delay costs are the consequences of interoperability problems on schedule and delivery.

3.2.1   Stakeholder Comments on Avoidance Costs

Nearly all of the architects and engineers and owners and operators interviewed cited inadequate current and legacy IT system connectivity as a problem. They subsequently incur significant bandwidth, training, and software maintenance labor charges. In addition, task-specific systems do not communicate well with one another.

Savings estimates in terms of personnel were projected by one national design firm. This design firm typically spends approximately $50 million per year for professional salaries, and a 10 percent increase in efficiency, possible with greater electronic interoperability, could result in $5 million in avoided direct-labor expenses.

A second A&E firm devotes internal research and development funds solely for the purpose of developing standards and integration tools between its engineering, construction, and procurement systems. Similarly, most of the larger organizations interviewed indicated that they participate in several industry consortia aimed at improving interoperability issues, thereby incurring additional labor charges and travel expenses.

Avoidance costs related to managing the paper trail of large-scale construction projects are high. For example, one general contractor noted that one skyscraper project had five full-time employees onsite managing the reams of paper designs, engineering specifications, and communication. A similar comment was provided for a large-scale coordinated construction effort in the late 1990s.

3.2.2 Stakeholder Comments on Mitigation Costs

Stakeholders indicated they often manually reenter data between systems and frequently transit back and forth between paper-based and electronic systems, incurring redundant labor costs. As a consequence, they must verify that all parties, no matter which phase the construction project is in, have the same information concerning designs, systems planning, and specifications.

One general contractor said that there are usually more than 200 requests for information during a typical project. Each request entails sifting through paper documents to locate the information needed. The same general contractor reported that he would like to use CAD more frequently onsite, but that CAD drawings show too much detail. That detail impedes the viewing of construction-phase specific information because certain layers of information imbedded in the file cannot be removed to show only what is necessary to field personnel.

Echoing the general contractor's information request comments, a major manufacturing owner and operator conducted internal studies and determined that typical design, construction, operations, and maintenance engineers spend 40 to 60 percent of their time looking for information and validating it. An owner and operator with a large commercial real estate portfolio indicated that the typical onsite building engineer spends upward of 15 percent of each day simply tracking down information to handle maintenance requests.

One facilities expert interviewed said that every dollar saved connecting the design to construction would generate savings in an amount 10 times more when connecting the operations and maintenance controls to the original CAD and engineering analysis design. The expert believed that the downstream effects in operations and maintenance are more important than connecting design software to construction. A major manufacturer completed a study that identified a projected savings of $51 million annually on three plants alone. For a public utility conducting a similar study, the numbers were $32 million annually for a 400 megawatt power plant.

3.2.3 Stakeholder Comments on Delay Costs

Avoidance and mitigation costs impact the scheduling of construction and operations and maintenance activities; delay costs are their consequence. Stakeholders cited late penalties, time delays, and idle resources as key delay costs. One manufacturing owner and operator also estimated that they could increase their production line "up-time" by 2 to 3 percent if they achieved true system interoperability, thereby increasing return on investment. They estimated that this equates to millions of dollars for their company annually.

The insights from the preliminary interviews informed the methodology and economic and technical impact metrics presented in the following two chapters. Chapter 7 revisits stakeholders' views of interoperability, in particular their views on the challenges and impediments to improved interoperability and connectivity opportunities.

4 Methodology for Quantifying Interoperability Costs

This chapter builds on the background information and initial analysis presented in previous chapters and describes the economic analysis framework with which the costs of inadequate interoperability in the capital facilities industry can be measured. The methodology includes a description of the technical and economic metrics for quantifying costs and methods for extrapolating survey information to develop national impact estimates.

The cost estimate of inadequate interoperability was quantified by comparing the current state of interoperability with a hypothetical counterfactual scenario in which electronic data exchange and availability is fluid and seamless. The difference between the current and counterfactual scenarios represents the total estimated economic loss associated with inadequate interoperability. Costs were calculated at the social level. In other words, this analysis quantified the efficiency loss borne by society because of inadequate interoperability.

4.1 INTEROPERABILITY COST CHARACTERISTICS

The estimation approach focuses on identifying and quantifying the interoperability *efficiency loss* associated with construction-related activities. During the interviews, *opportunity losses* associated with interoperability problems were also investigated, but these costs were not included in the quantitative analysis because of their speculative nature. The analysis approach aimed to estimate costs that could be reliably documented, realizing that the results are likely to underestimate total interoperability costs.

In the context of this analysis, three general activity cost categories were used to characterize inadequate interoperability: avoidance costs, mitigation costs, and delay costs.

Avoidance costs are related to the ex-ante activities stakeholders undertake to prevent or minimize the impact of technical interoperability problems before they occur. Examples include

- the cost of purchasing, maintaining, and training for redundant CAD/CAE systems;

- the cost of maintaining redundant paper systems for exchanging information;

- outsourcing translation services to third parties;

- investments in in-house programs, such as point-to-point translators and neutral file format translators to address interoperability issues; and

- the cost of participating in industry consortia activities aimed at improving interoperability.

Mitigation costs stem from ex-post activities responding to interoperability problems. These are often the largest portion of interoperability costs (Martin and Brunnermeier, 1999; Gallaher, O'Connor, and Phelps, 2002). Most mitigation costs result from electronic or paper files that have to be reentered manually into multiple systems and from searching paper archives. Mitigation costs in this analysis may also stem from redundant construction activities, including scrapped materials costs. In summary, mitigation costs generally include

- the cost of design and construction rework due to interoperability problems,

- the cost of manually reentering data when electronic data exchange is unavailable or when errors were made in the exchange, and

- the cost of verifying information when original sources cannot be accessed.

Delay costs arise from interoperability problems that, for example, delay the completion of a project or increase the length of time a facility is not in normal operation. These costs are the most difficult to quantify and include

- idle resources as construction activities are delayed,

- profits lost due to delay of revenues (discounts the value of future profits),

- losses to customers and consumers due to delay in the availability of products and services, and

- idle resources when a facility is not in normal operation.[15]

Industry stakeholders are typically well aware of their delay costs in terms of project delays or facility down-time. The key to estimating delay costs is determining where the bottlenecks are and which data exchange activities are timeline critical and which are not.

When investigating avoidance, mitigation, and delay costs, it is important to distinguish between the economic impact on stakeholder groups and the impact on U.S. social welfare. For example, penalties assessed due to delays are primarily transfer payments between stakeholder groups, and they impact the distribution of wealth but not necessarily total social welfare. These penalties are a measure of private costs; the extent to which private costs reflect social costs depends on market conditions. For example, in a perfectly competitive market, other firms will increase their output to meet demand if delays or down-time limit one firm's production. This leads to a redistribution of revenue but minimal social costs. In contrast, delays in the availability of unique or enhanced products and services directly lower social welfare and can impact construction stakeholder groups and consumers.

4.2 DEFINITION OF CAPITAL FACILITIES INDUSTRY SCOPE

The wide scope of U.S. construction activity impedes collection of comprehensive, industry-wide interoperability data within the time and resource constraints of this analysis. Therefore, the study's scope is limited to the capital facilities industry, whose sophisticated information requirements, discussed in Chapter 3, generate the majority of data exchange activity in the construction industry. The capital facilities industry covers activities related to all stages of commercial-buildings and industrial-facilities life-cycle management. Consequently, selected

---

[15]Late penalties are not included as a delay cost as they are not economic losses, but transfers between stakeholder groups.

architecture and engineering disciplines are included in our analysis, in addition to general contractors and owners and operators.

4.3 MODELING APPROACH

The cost of inadequate interoperability was quantified by comparing the current state of interoperability with a hypothetical counterfactual scenario in which electronic data exchange and availability is fluid and seamless. The concept of fluid and seamless data management encompasses all process data directly related to the construction and facility management process, including initial designs, procurement information, as-builts, and engineering specifications for operations and management. The difference between the current and counterfactual scenarios represents the total economic loss associated with inadequate interoperability.

4.3.1 Development of the Counterfactual Scenario

As stated, this analysis compares the current status of construction information management to a hypothetical scenario in which interoperability issues do not occur. The use of counterfactual analysis, pioneered by Robert Fogel and once extensively debated, has become well accepted over the past 20 years (Fogel, 1979).

The specification of the counterfactual scenario strongly influences the calculated economic cost of inadequate interoperability. Admittedly, the construction of a counterfactual scenario is a synthetic exercise; it is difficult or impossible to fully describe with a high degree of confidence a situation that does not exist. For this reason, development of the counterfactual scenario entailed discussions with a wide range of stakeholders throughout the capital facility industry.

For this analysis, the counterfactual scenario is defined as a world where the electronic exchange, storage, and retrieval of building blueprints, configurations, business data, and engineering specifications are seamless. Stakeholders in each stage of the construction life cycle would have ready access to electronic information using information technology equipment, including computers and handheld devices.

In the counterfactual scenario, information would be available to all stakeholders and their employees when the information is required. This implies that information needs be entered into electronic systems only once, after which it is available to relevant stakeholders instantaneously or on an as-needed basis through information technology networks and systems that are interoperable and that make full use of standardization tools.

The current paradigm of limited and error-prone electronic data exchange and paper-based information management is then compared to a scenario in which the sharing of standardized electronic information is the norm. In this way, the analysis estimates the full potential benefit of seamless electronic design and data management relative to existing practices. The goal is to determine total interoperability costs in the industry, not necessarily the feasibility of attaining this level of industry standardization and information technology sophistication.

When quantifying the benefits (cost reductions) of interoperability, the focus of the study is on the timing, cost, and increased availability of currently collected information. The definition of the counterfactual world does not include increases in the accuracy or quantity of data collected. The focus is on the changes in business activities and costs associated with data availability — holding data quality constant. While it is true that improved interoperability will increase the value and hence the demand for improved and expanded data collection activities, these potential benefits (also referred to as "opportunity costs") will be investigated qualitatively and are not included in the empirical impact estimates.

Also, note that the interoperability improvements reflected in the counterfactual scenario will not eliminate all IT costs; incremental cost savings are primarily associated with removal of redundant hardware, software, and labor costs and improved business efficiency from increased data access.[16] Similarly, whereas significant cost reductions may result from the reduction of paper systems, some use of paper drawings may still be practical.

4.3.2 Time Frame of Economic Costs Estimation

The economic costs of inadequate interoperability were quantified for a calendar year. Given the complexity of developing retrospective and prospective cost flows, this analysis takes the approach of evaluating the current-year costs of inadequate interoperability (as opposed to costs

---

[16]Note that IT costs may actually increase if the use of electronic systems for managing and exchanging information increases substantially. Implementation of these systems would proceed only if the benefit (avoided interoperability costs) outweighed the costs (software and IT administrative support). This highlights that this study only estimates the costs on inadequate interoperability and does not investigate the cost of implementing interoperability solutions.

incurred throughout the lifetime of an individual project). Therefore, the analysis results are a "snapshot" of current efficiency losses in the capital facilities industry.

4.4 INADEQUATE INTEROPERABILITY COST ESTIMATION APPROACH

The estimation approach quantifies annualized costs that reflect interoperability problems throughout the construction life cycle. Construction projects and facility operations are segmented into four life-cycle phases. Interoperability problems affect an array of stakeholders and encompass a large number of activities. Thus, the estimation approach is built on a 3-D framework:

- *Facility Life Cycle*: Planning, design, and engineering; construction; operations and maintenance; and decommissioning.

- *Stakeholder Groups:* Aggregated to architects and engineers (A&Es), general contractors (GCs), specialty fabricators and suppliers (SFs), and owners and operators (OOs).

- *Activities Categories*: Efficiency losses from activities incurring avoidance, mitigation, and delay costs.

This approach quantifies inadequate interoperability costs for each stakeholder's activity category during each life-cycle phase. The approach begins by separating the life cycle into four phases, each of which includes the range of activities for each stakeholder and for which the exchange, use, or manipulation of electronic information is relevant. Technical impact metrics are identified for each activity and paired with an appropriate economic metric to estimate costs.

The estimation approach can be represented as a 3-D framework (see Figure 4-1). The first dimension identifies the four life-cycle phases and the second dimension identifies each stakeholder group for which costs were quantified. In these two dimensions, the rectangle in the figure presents the sum of interoperability costs for each stakeholder during each phase of the life cycle. However, in the process of deriving the costs, the first step was to estimate efficiency losses by activity category (for each life-cycle and stakeholder group). This disaggregated estimation approach, where each individual cell within the matrix was quantified, allowed flexibility in the presentation of results in the following chapters.

Figure 4-1. 3-D Representation of Estimation Approach of Inadequate Interoperability Costs

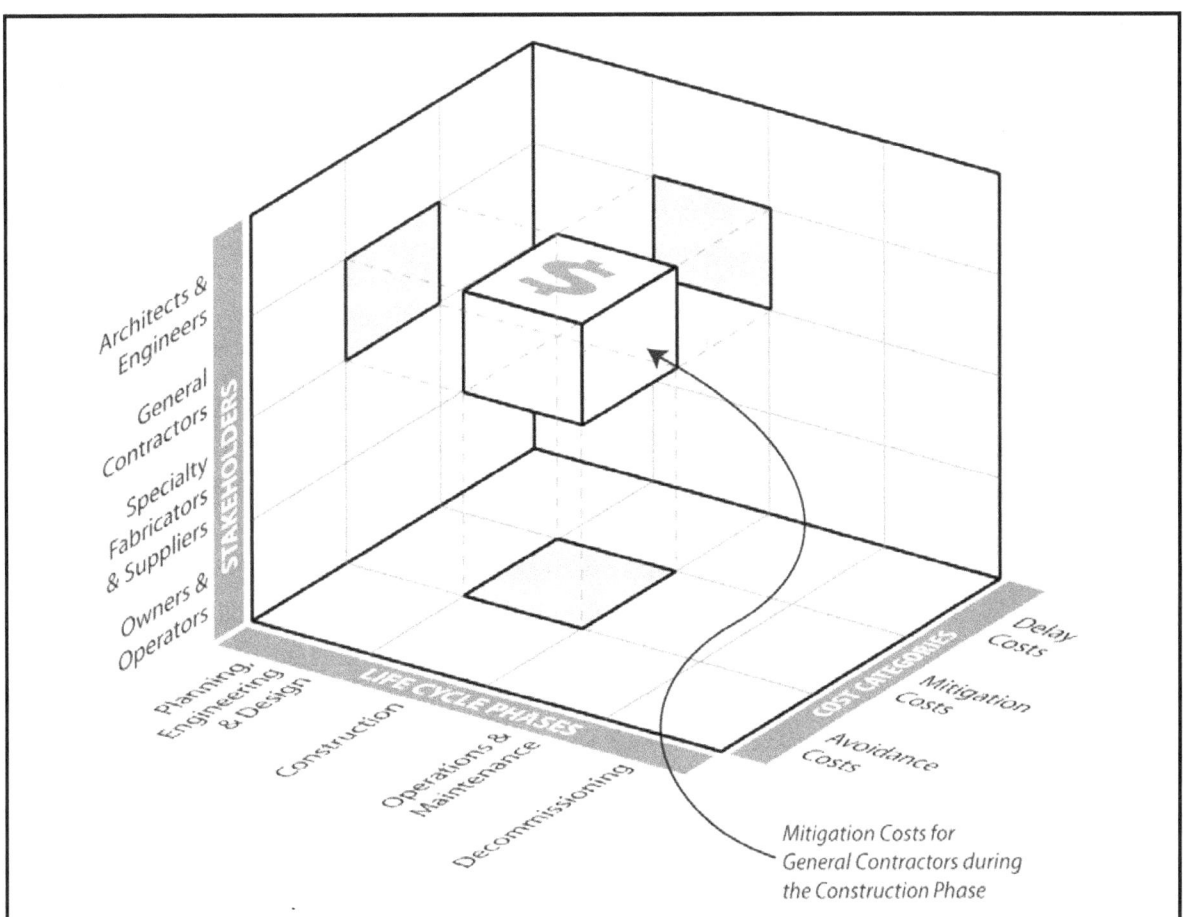

4.4.1 Interoperability Costs by Life-Cycle Phase

The facility life-cycle is segmented into four phases. Each phase includes multiple industry stakeholder groups and activities that are affected by interoperability problems. The life-cycle phases used in the analysis are as follows:

- *Planning, design, and engineering.* Includes all activities that occur prior to the construction of a new facility. Costs are typically one-time costs and are incurred by all stakeholders within the capital facilities industry.

- *Construction.* Includes all activities that occur during the building of a new facility. These are typically one-time costs associated with initial facility construction. However, the construction phase also includes major renovations and remodeling work. Interoperability costs affect all stakeholder groups but are concentrated on general contractors and specialty fabricators and suppliers.

- *Operation and maintenance.* Includes ongoing annual activities related to performing routine and as-needed maintenance for facilities during their operation. This phase also includes down-time and delay-cost impacts and costs are borne primarily by owners and operators.

- *Decommissioning.* Includes one-time interoperability costs general contractors and owners and operators incur due to lack of structural and materials information.

4.4.2 Technical and Economic Impact Metrics by Activity

Technical and economic impact metrics were used to quantify interoperability costs. Each technical and economic metric pair corresponds to a cost source or activity for which the exchange, use, or manipulation of electronic information is impacted. These activities are grouped as cost sources pursuant to the avoidance, mitigation, and delay cost paradigm. For example, administrative staffing for exchanging paper diagrams and designs is categorized as an avoidance cost because these activities are conducted to avoid problems associated with the exchange of electronic data. In contrast, data reentry and information verification are categorized as mitigation costs because they are follow-up activities to correct interoperability problems associated with the exchange of electronic data.

Technical impact metrics were used to measure the labor activities, and capital and material inputs resulting from interoperability problems. These commonly include wasted time or decreased labor productivity, measured in terms of labor hours and unnecessary or redundant software and computer systems measured in terms of the number of software applications or licenses (seats).

Economic metrics were then used to value the technical impacts in terms of labor costs (wage rate multiplied by hours) and expenditures for materials (price multiplied by quantity). Economic impact metrics provided the means by which the technical metrics were translated into economic impact.

Table 4-1 presents technical and economic impact metrics grouped by interoperability cost category. The metrics were identified and refined during scoping interviews with industry stakeholders and provide structure for the survey instruments. As discussed in Chapter 5, surveys were primarily used to obtain information on technical impact metrics.

Table 4-1. Summary of Technical and Economic Impact Metrics

Source of Cost	Impact	Technical Metric	Economic Metric
Avoidance Costs			
Redundant CAD/CAE systems	CAD/CAE software licenses	Number of software licenses required by type	Expenditures for software licenses
	System maintenance	Labor required to maintain all software systems	Cost of labor required to maintain software systems
	System training	Labor hours devoted to training and gaining competence on all systems	Cost of labor time required to gain competence on all systems
	Productivity loss	Lost productivity and time spent on secondary systems	Value of labor resources lost
Multiple translators	Translation software licenses	Number of translation software licenses required by type	Expenditures for translation software licenses
	Software training	Labor hours devoted to training on the use of different translators	Cost of training labor to use different translators
Paper systems	Productivity loss	Lost productivity associated with maintaining paper-based communication systems	Value of labor resources lost
Outsourcing data translation	Third-party suppliers	Jobs outsourced to third-party suppliers of data exchange services	Cost of outsourced work
Investments in interoperability solutions	In-house interoperability research	Capital, labor, and materials devoted to in-house interoperability research	Cost of in-house interoperability research
	Activities in industry consortia	Time and materials devoted to participation in industry consortia	Cost of membership, labor time, and materials devoted to consortia activities
Mitigation Costs			
Scrapped efforts	Design changes due to inadequate information access	Hours required to rework designs	Cost of time required to rework designs
	Construction changes due to inadequate information access	Labor hours lost and scrapped construction material	Value of labor resources lost and scrapped construction material

(continued)

Table 4-1. Summary of Technical and Economic Impact Metrics (continued)

Source of Cost	Impact	Technical Metric	Economic Metric
Mitigation Costs (continued)			
Inadequate information sharing	Manual data reentry: electronic information sources	Number of jobs that required reentry of information from electronic sources and the average time per job	Value of labor resources lost
	Manual data reentry: paper-based information sources	Number of jobs that required reentry of information from paper-based communication and the average time per job	Value of labor resources lost
	Productivity loss	Lost productivity associated with searching for, providing, and validating paper-based information	Value of labor resources lost
Delay Costs			
Delays	Delayed products and services	Length of delay and the productive capacity of the facility	Length of delay times the value of the enhanced product or service per period of delay
	Delayed profits	Length of delay and the quantity of products or services that would have been sold per period of delay	Value of profits with no delay less value of profits discounted over period of delay
	Idle resources	Incremental labor hours incurred and materials lost due to idled facility	Value of labor and materials resources lost

The paired economic metrics were mostly obtained from secondary information sources, such as the Bureau of Labor Statistics's Occupational Employment Survey for average wage rates by labor category and market data for software and material prices.

4.4.3 Interoperability Cost Crosswalk between Life-Cycle Phase and Stakeholders

To fully understand interoperability issues and market barriers to the development and adoption of interoperability solutions, it is important to estimate the distribution of interoperability costs across stakeholders.

The estimation approach disaggregates interoperability costs by life-cycle phase and activity by four stakeholder groups.

Table 4-2 presents the hypothesized crosswalk between stakeholders and cost categories along with a general indicator about the magnitude of costs for each stakeholder by life-cycle phase. Per-unit cost impacts were estimated for each cell and aggregate impacts were calculated by using secondary population data to weight the per-unit impact estimate.

Table 4-2. Hypothesized Distribution of Interoperability Cost by Life-Cycle Phase, Stakeholder Group, and Activity Category

Life-Cycle Phase	Stakeholder Group	Cost Category		
		Avoidance	Mitigation	Delay
Planning, Design, and Engineering	Architects and engineers	●	●	
	General contractors	●	○	
	Specialty fabricators and suppliers	○	○	
	Owners and operators	●	●	
Construction	Architects and engineers	○	○	
	General contractors	●	●	●
	Specialty fabricators and suppliers	●	○	○
	Owners and operators	●	●	●
Operations and Maintenance	Architects and engineers	○		
	General contractors	○		
	Specialty fabricators and suppliers			
	Owners and operators	●	●	●
Decommissioning	Architects and engineers			
	General contractors		●	○
	Specialty fabricators and suppliers			
	Owners and operators		●	○

Note: ○ = some interoperability costs and ● = significant interoperability costs.

As indicated in Table 4-2, not all stakeholder groups have costs in each life-cycle phase. Therefore, each stakeholder group was administered a different survey to reduce the information collection burden. For example, architects likely would have significant avoidance costs from supporting multiple CAD systems and mitigation costs from imperfect data exchanges. In contrast, general contractors may have less avoidance costs, because they generally use less information technology, but would have significant mitigation costs.

4.5 GENERATING NATIONAL-LEVEL IMPACT ESTIMATES

National-level impact estimates were generated by extrapolating survey responses using secondary information sources. The extrapolation plan used the existing stock (square feet) and annual growth (square feet per year) of capital facilities. The following chapter describes the building types included in the stock and flow square footage totals. The discussion that follows presents the extrapolation method for applying these weights.

4.5.1 Plan for National Impacts by Stakeholder Group and Life-Cycle Phase

Survey respondents were asked to provide the total square footage of construction or operating space associated with the interoperability costs they provided. For example, owner/operators of large commercial office buildings may have provided interoperability costs for their entire real estate portfolio. These costs were normalized by the total portfolio's square footage. The average cost per square foot across all survey respondents for a specific cell (as shown in Figure 4-1) was then weighted by the national square footage to estimate national interoperability costs.

Costs per square foot were aggregated to the life-cycle level and then weighted. One-time activities were weighted by square footage growth (decline) rates and ongoing (annual) activities were weighted by the cumulative stock of existing square footage. Simplified algebraically, the total interoperability costs in a single year can be expressed as:

$$\text{Cost} = \sum_{ij} (DEP_{ij} {}^* Qg) + \sum_i (C_{ij} {}^* Qg) + \sum_i (OM_{ij} {}^* Qs) + \sum_i (D_{ij} {}^* Qd),$$

where

i = the stakeholder group subscript;

j = the activity category subscript;

DEP_{ij} = annual design, engineering, and planning interoperability cost per square foot or capacity for stakeholder i for activity j;

C_{ij} = annual construction interoperability cost per square foot or capacity for stakeholder i for activity j;

OM_{ij} = annual operations and maintenance interoperability cost per square foot or capacity for stakeholder i for activity j;

D_{ij} = annual decommissioning interoperability cost per square foot or capacity for stakeholder i for activity j;

Qg = total capital facility square footage under construction in a given year;

Qs = total capital facility existing stock in terms of square footage; and

Qd = total capital facility square footage decommissioned in a given year.

Note that DEP_{ij}, C_{ij}, OM_{ij}, and D_{ij} represent the cost-per-square-foot impact estimates for an individual stakeholder group's activity category. Industry surveys were used to collect and estimate these per-unit interoperability costs; secondary data were used to estimate the weights. Note that Qs reflects the cumulative stock of facilities and will be significantly larger than Qg or Qd because they represent positive and negative flows in square footage, respectively.

5   Estimation Procedures and Data Sources

Chapter 4 presented the economic methodology for quantifying the costs of inadequate interoperability in the capital facilities supply chain. This chapter complements Chapter 4 by presenting the data sources and estimation procedures that underlie the quantitative results to be presented in the following chapter.

The estimation methodology integrates data collected from a variety of secondary sources with primary data collected via focus groups, telephone interviews, on-site interviews, and an Internet survey. To gather quantifiable interoperability cost data, each respondent completed a questionnaire that was specifically tailored to his or her stakeholder group. These surveys can be found in Appendix B to this report. Individual responses by stakeholder were aggregated and then extrapolated to the national level using the square footage data published by Energy Information Administration (EIA) surveys.

The purpose of this chapter is to

- present the secondary data sources employed by the analysis,

- describe how stakeholders' costs are distributed across capital facility life cycles,

- detail how each cost subcategory is defined and calculated, and

- describe how costs were extrapolated to the national level for each stakeholder group.

5.1 SECONDARY DATA SOURCES

The data gathered from these surveys were coupled with a variety of secondary data sources to minimize respondent burden, enhance the comparability of individual responses, and support the repeatability of results. The four main categories of secondary data employed in this analysis are

- employee wage rates,
- CAx and facilities management systems annual maintenance costs,
- national square footage estimates of the total existing stock, and
- new construction flow of capital facilities.

The latter two bullet points will be presented at the end of this chapter with the extrapolation methodology.

5.1.1 Wage Estimates

All wage estimates employed in this analysis are from the national Occupational Employment Statistics (OES) for 2002 as compiled by the Bureau of Labor Statistics (BLS). The OES provides national average hourly wage estimates for many occupations at both the national and industry-specific levels.

BLS wage rates were used to quantify the hourly productivity benefits and labor savings. To monetize these benefits categories, the labor hour savings were multiplied by the appropriate wage rate for the employee functioning in that position. To simplify discussions in later sections, the employment categories and wage rates used in these calculations are presented and discussed here.

The BLS tracks wages by job category and by industry. As is the case for most positions, wage rates for similar positions vary by industry according to each industry's supply and demand for that labor. The following portions of the BLS OES data were used to gather stakeholder-specific wage estimates:

- Architects and Engineers: NAICS 541300—Architectural, Engineering, and Related Services
- General Contractors: NAICS 236200—Nonresidential Building Construction
- Specialty Fabricators and Suppliers: NAICS 238200—Building Equipment Contractors

- Owners and Operators: 2002 National Occupational Employment and Wage Estimates

National wage estimates were used for owners and operators because this stakeholder group spans all segments of the U.S. organizational landscape.

Table 5-1 presents some frequently used wage rates employed in the discussion of estimation procedures. The entire set of wage rates and labor categories used in calculating inadequate interoperability costs is listed in Appendix C. The term "CAx user" is generically applied to those employees who deal directly with electronic and paper-based information and design problems. CAx is an abbreviation for CAD, CAE, and CAM systems. The original BLS data have been multiplied by a factor of 2.0 to estimate the fully loaded wage rates that include employee benefits, such as employer-sponsored health and dental insurance and 401(k) contributions, and administrative and overhead costs, such as facilities and equipment expenses.

Table 5-1. Key Wage Rates Employed to Quantify Costs of Inadequate Interoperability

Stakeholder Group	BLS Occupation Title	Mean Hourly Wage (2002)	Estimated Loaded Hourly Wage
Architects and Engineers			
CAx user (architecture)	Architects, except landscape and naval	$29.88	$59.76
CAx user (engineering)	Civil engineer	$30.53	$61.06
General Contractors			
CAx user	Civil engineer	$28.57	$57.14
Construction laborer	Construction laborer	$14.72	$29.44
Specialty Fabricators and Suppliers			
CAx user	Civil engineer	$27.20	$54.40
Construction laborer	Construction laborer	$14.01	$28.02
Owners and Operators			
CAx user	Architects, except landscape and naval	$30.06	$60.12
Operations and maintenance engineer	Civil engineering technician	$18.71	$37.42

Source: BLS, 2004.

5.1.2 Annual Maintenance Costs

To exchange data or review incoming or outgoing electronic files, firms often maintain software licenses for CAx systems that are not their in-house or primary software tools. They also invest in alternate CAx systems if they are awarded a contract that stipulates the use of a system different from their primary system. Using a combination of publicly available pricing schedules and informal interviews with software vendors, the approximate annual maintenance fees were obtained for a large number of CAx and facilities management system configurations. It is a common practice for vendors to offer discounts to customers based on the volume of licenses purchased. This practice was taken into account by associating each firm's number of licenses with the correct price for that volume of licenses. Thus, the annual fee applied per license for a particular system varied among respondents depending on the number of licenses they held.

To protect the confidentiality of participants and prevent the disclosure of vendors' proprietary annual licensing fee schedules, the annual software license and maintenance fees for CAx and facilities management systems are not disclosed by stakeholder group or software package. However, the median annual cost per license for all systems covered in this report, including discounts, was $690 in 2002.

5.2 DISTRIBUTING COSTS BY LIFE-CYCLE PHASE

Extensive on-site stakeholder interviews and focus groups provided a wealth of data to facilitate the distribution of some costs by life cycle. Avoidance costs in particular span all life-cycle phases because they involve investments to prevent interoperability problems from occurring. Collecting avoidance costs by life-cycle phase was not possible because firms were not able to specifically allocate avoidance activities solely to a particular life-cycle phase. In large part this is because avoidance costs consist of essential business infrastructure. Some measure for distributing avoidance costs by life-cycle phase was necessary. Thus, the percentage allocation of avoidance costs by life-cycle phase was a key area of investigation. Mitigation and delay costs were collected by life-cycle phase because firms could specify activities incurring cost for each phase. Therefore, only avoidance costs required measures for distribution by life-cycle phase.

The distribution measures presented in Table 5-2 were derived by seeking a consensus from on-site interviewees and focus group participants. The distribution measures were based on the facilities-related activity level for each stakeholder by life-cycle phase. Architects and engineers reported that the distribution of their activity level for a typical project was 85 percent during the planning, engineering, and design phase and 15 percent during the construction phase. General contractors and specialty fabricators and suppliers reported that the percentage distribution of their activity was approximately the inverse of that for architects and engineers. None of these three stakeholder groups indicated that they engaged in a significant amount of activity during the operations and maintenance (O&M) phase of a capital facility. Owners and operators reported a distribution of capital facilities-related activity that covered all three phases for which costs were quantified— 27.5 percent for planning, engineering and design; 22.5 percent for construction; and 50.0 percent for O&M.

Table 5-2. Percentage Distribution of Avoidance Costs by Life-Cycle Phase, by Stakeholder Group

Stakeholder Group	Planning, Engineering and Design Phase	Construction Phase	Operations and Maintenance Phase
Architects and Engineers	85.0	15.0	
General Contractors	15.0	85.0	
Specialty Fabricators and Suppliers	15.0	85.0	
Owners and Operators	27.5	22.5	50.0

Source: RTI estimates.

All the avoidance costs quantified in this report are distributed according to the proxies presented in Table 5-2, with one noted exception. Avoidance costs related to facilities management systems were allocated solely to the O&M phase. No owner and operator indicated that these systems were used during either the planning, engineering, and design phase or the construction phase.

5.3 ESTIMATING PROCEDURES FOR INTEROPERABILITY COSTS

The following presents the procedures for calculating avoidance, mitigation, and delay costs. Not all cost categories apply to each stakeholder. Therefore, the acronym for each stakeholder group for which the cost category is applicable follows the subheading for each estimation procedure.

It is important to note that each impact category employs data specifically related to inadequate interoperability. The survey instruments contained questions that requested only the potential reductions if perfect interoperability were achieved. Thus, it should not be inferred from this analysis that specific cost categories could potentially be eliminated on the whole. Rather costs could be reduced. For example, RFIs will continue to be shared among stakeholders; however, the volume of RFIs would likely be less. Costs incurred because of inherent design and/or construction flaws are also excluded by the questionnaires because they do not represent a direct interoperability problem. Such costs are only captured if they relate specifically to inadequate information management and exchange.

5.3.1 Calculating Avoidance Costs

Avoidance costs are those costs an organization incurs to prevent the occurrence of interoperability problems. The costs include those incurred to use and maintain redundant information technology systems. They also include the business processes, such as cost estimation and accounting, where costs exist due to the lack of adequate information management and exchange as defined by the counterfactual scenario. The following discussion illustrates how avoidance costs were estimated.

The acronym for each stakeholder group for which the cost category is applicable follows the subheading for each estimation procedure.

Licensing Costs for Redundant CAx Systems—A&E, GC, SF, OO

Licensing costs for redundant CAx systems, which are primarily computer-aided design (CAD) and computer-aided engineering (CAE) systems, were quantified by multiplying the annual per-license cost and the number of licenses for systems that respondents indicated were redundant. Systems were classified as redundant if the respondent indicated that they duplicated the in-house systems on which their organization was standardized. (Companies often maintain redundant systems to work with other firms that have standardized on a competing system.)

For example, suppose a firm maintains two systems, System A and System B, and prefers System A. If System B duplicates System A's functionality, then the costs associated with System B were considered redundant. If the firm has 100 licenses for System B, and each license cost $500 (taking into account discounts based on license volume), the calculation would be

> 100 (the number of licenses for System B)
>
> > x $500 (the annual maintenance and license cost per license for
> >
> > System B)
> >
> > = $50,000 in redundant CAx systems costs.

This analysis considers that original purchase and installation costs for redundant systems are nonrecurring costs and are therefore considered sunk costs. These costs are not included in this calculation. However, many firms are still in the process of investing in these systems. We reiterate that the estimate developed is a lower-bound estimate.

Productivity Losses and Training Costs for Redundant CAx Systems—A&E, GC, SF, OO

Most organizations standardize on one system on which their employees are trained and are accustomed to using in the performance of their duties. These organizations may also have other software products that tie into their primary systems and enhance these systems' functionality. However, if the firm maintains redundant systems, it suffers a productivity loss and training costs associated with these duplicative systems. Previous studies have indicated that users are 70 percent less productive when using redundant (secondary) systems due to infrequent usage (Gallaher, O'Connor, and Phelps, 2002). In addition to incurring productivity losses, organizations also invest in training on redundant systems, a cost they would otherwise not incur if seamless interoperability existed.

For example, suppose an architecture firm had a CAx user population of 100, and that 10 percent of those users spent 10 percent of their time working in redundant systems. As listed in Table 5-1, the loaded hourly wage for CAx users for architects is $59.76, and each calendar year has 2,000 work hours. Interoperability costs can be calculated as follows:

100 CAx users

> x 0.10 (proportion using redundant systems)
>
> x 0.10 (proportion of time working in those systems)
>
> x 0.30 (proportion of productivity lost)
>
> x $59.76 (average loaded hourly wage)
>
> x 2,000 (number of work hours in a year)
>
> = $35,856 (annual productivity loss)

To maintain competency in using those systems, CAx users undergo periodic training. However, total training costs were difficult to estimate because formal training is less common than "on the job" training. Therefore, it is difficult to distinguish between lost productivity and training-related costs.

Several organizations estimated that each user received on average about 40 hours of formal training on their secondary systems. Although this figure may be an underestimate, interviewees thought this was the best possible figure they were able to provide, given that they more rigorously track training-related expenses for their in-house or preferred systems.

Estimates on the tuition and fees paid to third-party training centers were unavailable. However, course fees are generally minimal compared to the labor expense of sending an employee for training; hence, tuition and fees were not included in this analysis.

Training costs were calculated using CAx user work life, the amount of time spent in training, redundant CAx systems user population, wage rates, and annual work hour estimates. The method for estimating annual training cost is illustrated below:

1. Each CAx user's work life was estimated to be 25 years. It was therefore assumed that, if a firm has 100 CAx users, about four employees per year will be new to the industry.

2. Over the course of that work life, it was estimated by participants that a user would receive 40 hours of formal training on redundant systems.

3. Not all new employees are trained on a redundant system. Therefore, it is necessary to multiply employee turnover by the percentage of users who work on redundant systems and the percentage of redundant systems licenses. If 10 percent of users work in redundant CAx systems, the annualized number of employees receiving this training would be 0.40.

4. If the user were an architect, then the 0.40 estimate would be multiplied by the training hours (40) and the wage rate ($59.76).

5. As a result, the firm's annualized expenditures for training on redundant systems would be $956.16.

Thus, the total annual productivity loss on redundant CAx systems and investments in redundant CAx systems training for this hypothetical architecture firm would be $36,812.16.

IT Support Staffing Costs for Redundant CAx Systems—A&E, GC, SF, OO

To maintain their software investment and to support employees using that software, organizations employ computer network and systems administrators, software support specialists, and design support specialists. These employees maintain smooth operation of networks and troubleshoot technical problems. Although IT infrastructure and software represent additional expenditures, these costs are minimal compared to the labor needed to support them and are therefore excluded from the analysis.

The relationship between a firm's number of CAx systems and its IT staffing employment is not linear. For example, IT staffing would not be cut in half if the firm reduced its number of systems by one half. For most firms, there is a base number of employees for one system and some incremental number of employees for each additional system. During the on-site interviews, respondents provided detailed information on the support staffing they retained and how that staffing would change if they were to eliminate their redundant CAx systems.

For example, one firm said that it had 500 CAx users. Approximately 39 IT staff would be needed to support the CAx users if all were working on primary systems. Yet 46 IT employees (about a 20 percent increase) are needed because two systems are being supported.[17] Similar responses, averaged across all on-site interviewees, were used to compute the average increase in IT staff relative to the number of CAx systems. Table 5-3 lists the aforementioned firm's current CAx IT staffing and indicates how the staffing would change with the addition of another system.

---

[17] If an additional system were added, but with only a small number of users, the impact would not be as great. As part of the analysis, the ratio of in-house systems to total licenses was used in the firm-level calculations to adjust the IT staffing increases accordingly.

Table 5-3. Sample Change in IT Staffing Associated with Redundant Systems

IT Positions	Staffing Level with 1 System	IT Staffing Level with 2 Systems
Computer and Network Systems Administrators	20	25
Design Support Specialists	6	7
Software Support Specialists	13	14
Total	**39**	**46**

Source: Survey participant.

Algorithms were developed to calculate redundant IT staffing costs as a function of several variables collected during the surveys and interviews. The large number of variables included in the calculation is attributable to the complex formula needed to estimate, for each respondent, how IT staffing would change if seamless interoperability were fully implemented and the number of CAx systems reduced.

The IT staffing calculations were based on the following information:

- number of CAx users;
- existing number of CAx systems;
- potential number of CAx systems;
- ratio of in-house system licenses to total licenses;
- annual work hours (2,000);
- wage rates for each of three IT staffing positions that were included in the analysis; and
- incremental staffing coefficients.

The calculations took into account the number of system reductions and calculated the difference between the estimated current number of IT professionals in each position and the number required under the counterfactual scenario. The resulting staff reduction was then multiplied by the annual number of work hours and by the appropriate wage rate for each position and industry.

Data Translation Costs—A&E, GC, SF, OO

Organizations were asked to provide the annual costs for internal translation tools and third-party translation services. Organizations often outsource interoperability problems to solutions providers. The counterfactual scenario indicates that this effort would be preempted by a world in which information management and exchange were seamless and electronic.

Research and Development Costs—A&E, GC, SF, OO

Research and development (R&D) costs are those costs incurred by organizations to participate in internal and consortia R&D efforts. Organizations were asked to provide the annual fees paid for membership in industry consortia. In addition, the total number of CAx user hours spent annually on internal R&D and participation in consortia activities were monetized using the loaded CAx user wage rates presented in Section 5.1.1.

Inefficient Business Process Management Costs— A&E, GC, SF, OO

All organizations have administrative and internal service functions that support their revenue centers. Interoperability affects these functions because information management and exchange issues increase the work load for these functional areas. This study therefore considers the effect of inadequate interoperability on internal business functions. Inefficiency or redundant business process management costs are considered avoidance costs because organizations maintain these staffing levels in recognition of current and future interoperability problems.

Redundant business process management costs occur when inadequate interoperability ripples through the organization, from revenue centers to administrative and technical support functions. The business processes considered are presented in Table 5-4. Not all processes were evaluated for each stakeholder; therefore, Table 5-4 indicates which processes are applicable to each stakeholder group.

Study participants were asked to estimate the total staffing (in full-time equivalent workers [FTEs]) for each business process, and the percentage staff reduction that could be achieved if information exchange and management were both seamless and electronic. The potential reduction was then multiplied by the appropriate wage rate for that business process and 2,000 annual work hours per employee. Due to the large number of wage rate estimates employed in costing redundant business processes, these wage rate estimates are presented, by stakeholder group, in Appendix C. The impact estimates presented in the following chapter will present the total cost of inefficient business process management for each stakeholder, by life-cycle phase.

Table 5-4. Business Processes Impacted by Inadequate Interoperability, by Stakeholder Group

Business Processes	Architects and Engineers	General Contractors	Specialty Fabricators and Suppliers	Owners and Operators
Accounting	✓	✓	✓	✓
Cost Estimation	✓	✓	✓	✓
Document Management	✓	✓	✓	✓
Enterprise Resource Planning	✓	✓	✓	✓
Facility Planning and Scheduling	✓	✓	✓	✓
Facility Simulation	✓	✓	✓	✓
Information Request Processing	✓	✓	✓	✓
Inspection and Certification		✓	✓	✓
Maintenance Planning and Management		✓	✓	✓
Materials Management	✓	✓	✓	✓
Procurement	✓	✓	✓	✓
Product Data Management		✓	✓	✓
Project Management	✓	✓	✓	✓
Start-up and Commissioning		✓	✓	✓

Redundant Facility Management Systems Costs—OO

Redundant facilities management systems were calculated for owners and operators in a manner similar to that for the redundant CAx systems. This measure calculates the economic loss associated with instances in which firms or divisions within firms purchase facilities management systems that duplicate one another's capability. The total number of redundant systems licenses was multiplied by the average annual maintenance cost per license, taking into consideration any discounts based on license volume. The product was the total redundant facilities management systems costs. These costs applied only to the O&M life-cycle phase.

Productivity Loss and Training Costs for Redundant Facility Management Systems—OO

Productivity lost due to redundant facilities management systems and investments in redundant facilities management systems training were calculated using the same estimation procedures used for redundant CAx systems. However, the loaded wage rate applied was for O&M engineers. This cost applied only to the O&M life-cycle phase.

IT Support Staffing Costs for Redundant Facility Management Systems—OO

Information technology (IT) staffing costs for redundant facilities management systems were calculated for owners and operators in a fashion similar to those for CAx systems. Only two positions were considered for this cost area: network and computer systems administrators and software support specialists. The design support IT function was eliminated from this calculation because it is not applicable to this functional area. The coefficients presented for calculating CAx IT staffing changes were assumed to be the same for facilities management IT staffing because the employees were within the same department as those supporting CAx systems. This cost applied only to the O&M life-cycle phase.

5.3.2 Calculating Mitigation Costs

Mitigation costs are those costs incurred by organizations to resolve inadequate interoperability problems after they have occurred. They involve manual reentry of information, verifying that users are proceeding with the correct files, and rework due to proceeding with incorrect information obtained through inadequate information management and exchange. The surveys captured data specific to each life-cycle phase; mitigation costs are not distributed using weights in Table 5-2, as were the avoidance costs.

Costs of Manual Reentry—A&E, GC, SF, OO

Manual reentry costs are those costs related to the manual reentry of information between electronic systems, paper systems, and electronic and paper systems. Therefore, these costs are labor charges associated with correcting or reinputting data after an inadequate transfer. For manual reentry, the employee is replicating work that had already been completed by another, either within the firm or at a different organization.

These costs are incurred in the planning, engineering, and design and construction phases.

Respondents provided data on the number of hours CAx users spent each month performing manual reentry. To derive annual figures, the number of hours was multiplied by 12. This figure was then attributed a dollar value by multiplying it by the loaded hourly wage rate for CAx users.

Costs of Verifying Design and Construction Information—A&E, GC, SF, OO

Information verification costs are incurred when CAx users need to verify that they are working with the correct version of either paper or electronic files when multiple versions exist. Stakeholders in the capital facilities industry exchange design and construction files frequently; they indicated that the amount of time CAx users spend verifying that they are proceeding with the correct file is measurable. Therefore, this analysis quantified that cost by taking the monthly number of hours spent on information verification, multiplying it by 12 to derive annual hour estimates, and then multiplying that result by the appropriate loaded CAx user wage rate for each stakeholder. This calculation applies to the planning, engineering, and design and construction phases.

Costs of Reworking Design Files—A&E

Architects and engineers must rework design and engineering files when they proceeded with the incorrect versions of those files. This is a frequent occurrence in an industry in which multiple versions of the same files are present internally and at partner organizations. Respondents provided the number of hours they spent in an average month performing file rework. This number was multiplied by 12 and then by the CAx user rate to monetize costs. This calculation applies to the planning, engineering, and design and construction phases.

Costs of Post-Construction Redundant Information Transfer—A&E, GC

Architects and engineers and general contractors often perform redundant tasks when transferring files to owners and operators when construction ends and facilities have entered into service. The annual number of hours spent performing redundant tasks was multiplied by the CAx user wage to quantify costs. This calculation applies to the O&M phase.

Costs of RFI Management—A&E, GC, SF, OO

Under the counterfactual scenario, the number of requests for information (RFIs) issued would be far less than in the current scenario because of seamless electronic exchange and management of capital facilities data and information. Stakeholders spend a large number of hours responding to RFIs. This time was quantified by multiplying together the number of RFIs, the average number of hours spent responding to an RFI, and the appropriate CAx user rate for each stakeholder group. This calculation applies to the planning, engineering, and design and construction phases.

Costs of Construction Site Rework—GC, SF

Much like architects and engineers proceeding with the incorrect file version, general contractors and specialty fabricators and suppliers often proceed with incorrect design and engineering plans at job sites. When errors are uncovered, work must be redone and some materials are scrapped in the process. To quantify these costs, the survey instrument requested the number of times respondents performed construction site rework. They were asked to provide the number of labor hours spent redoing the work and the value of the materials scrapped. To estimate the total cost of performing rework at the job site, the number of construction laborer hours devoted to rework was multiplied by the loaded laborer wage rate, and the product was added to the value of the scrapped materials. This calculation applies only to the construction phase.

Costs of O&M Staff Productivity Loss—OO

O&M staff spent a significant amount of time tracking down information needed to perform maintenance and repair activities. During this time, they are less productive because they spend a measurable amount of time searching for and transferring information from a variety of electronic and paper-based information sources. This activity impacts their productivity; their time could better be spent resolving the maintenance or repair issue.

Owner and operator respondents provided the average amount of time O&M engineers spent transferring information between sources, a practice which adversely affected their productivity. For example, suppose a large owner and operator has 150 O&M engineers and that these engineers spend about 2 percent of their time performing redundant information transfers. Therefore, the productivity loss of this activity would be

> 150 O&M engineers
>> x 0.02 (proportion of time transferring information)
>> x $37.42 (average loaded hourly wage)
>> x 2,000 (number of work hours in a year)
>> = $224,520 (productivity loss)

This cost applied only to the O&M life-cycle phase.

Costs of O&M Information Verification—OO

O&M information verification costs were calculated using the same methodology to estimate information verification costs related to the first two life-cycle phases. The principal difference is that the wage rate used was the one for O&M engineers. This cost applied only to the O&M life-cycle phase.

Costs of O&M Staff Rework—OO

This calculation is similar to the construction site rework calculation for general contractors and specialty fabricators and suppliers. The only difference is that the wage rate applied was for O&M engineers and that this cost only applied to the O&M life-cycle phase.

5.3.3 Calculating Delay Costs

Delay costs are generally difficult to quantify because most firms plan for and accommodate delays in their project scheduling activities. However, in this study, it was possible to quantify delay costs, principally labor charges for idled employees. There are instances, particularly for general contractors, specialty fabricators and suppliers, and owners and operators, when construction laborers or O&M engineers are idled while waiting for information management and/or exchange issues to be worked out.

For owners and operators, delay costs were quantified for O&M engineers who spent a portion of their work days waiting for information needed to perform routine maintenance and repairs. These costs for owners and operators were quantified for the O&M life-cycle phase by multiplying the

- percentage of their time spent waiting for required information and not performing other tasks,
- annual work hours,

- number of O&M engineers on staff, and

- loaded wage rate for O&M engineers.

The estimation procedures are similar for construction laborers employed by general contractors and specialty fabricators and suppliers.

5.4 GENERATING NATIONAL IMPACT ESTIMATES

All calculated costs of inadequate interoperability were extrapolated to generate national impact estimates using square feet of floor space.[18] (The activity measure was square feet because national estimates for the stock and flow of capital facilities were available from survey data collected by the EIA.) The cost estimates derived from participant responses in each stakeholder group were aggregated and subsequently divided by the sum of their total activity.

5.4.1 Energy Information Administration Capital Facilities Floor Space Data

Data collected by two EIA surveys—the Manufacturing Energy Consumption Survey (MECS) and the Commercial Buildings Energy Consumption Survey (CBECS)—were used to extrapolate cost estimates by life-cycle phase for each stakeholder group. The MECS is conducted every 4 years and covers all industrial establishments in the United States. The most recent two surveys with available information collected data for 1994 and 1998. The CBECS is also conducted every 4 years. The most recent two surveys with available information collected data for 1995 and 1999. Because information was not available for 2002, the year for which inadequate interoperability costs were quantified, a series of adjustments were performed on the data prior to the extrapolation procedure.

MECS data provide national estimates of the total floor space for industrial, petrochemical, and utility facilities. To derive estimates of the floor space for these facilities, it was necessary to take the average annual floor space growth between 1994 and 1998 and use the same measure to develop the estimated total floor space in place for 2002. The 1994 and 1998 MECS indicated 12.329 billion square feet and 12.836

---

[18]It was preferred to use capacity data for industrial facilities. However, firms are reticent to provide detailed information on their facility capacity. In addition, the facility information that firms may choose to provide may not be comparable to national capacity estimates because of unit differences. Therefore, this analysis used square footage estimates for industrial facilities and commercial and institutional facilities.

billion square feet, respectively (EIA, 1997; EIA, 2001b). Because of information nondisclosure requirements, public use files were not available to screen the sampled facilities in the data set by facility size. Therefore, the entire facility stock figures provided by MECS were used.

The average annual floor space growth between 1994 and 1998 was 126.75 million square feet. This figure represents the best available estimate of the annual construction activity in square feet of these types of facilities. Therefore, 126.75 million square feet was used as the annual flow for industrial, petrochemical, and utility facilities. The estimate for 2002 was generated by adding the 1998 stock to four times the amount of annual flow, yielding an estimate of 13.343 billion square feet.

Unlike MECS, CBECS has public use data files available. Commercial and institutional facilities in the CBECS data set were filtered by floor space. Only those facilities, and their representative weights, with floor space equal to or greater than 100,000 square feet were included. The rationale behind this adjustment was to screen out small commercial establishments that were unlikely to have significant interoperability costs to participate in the analysis. After screening the 1995 and 1999 CBECS data, the total annual floor space was estimated to be 18.234 billion square feet and 22.276 billion square feet, respectively (EIA, 1998; EIA, 2002).

The average annual floor space growth between 1995 and 1999 was 1.010 billion square feet. This figure represented the best available estimate of the annual construction activity in square feet for these types of facilities. Therefore, 1.010 billion square feet was used as the annual flow for commercial and industrial. The estimate for 2002 was generated by adding the 1999 stock to three times the amount of flow, yielding 25.307 billion square feet for commercial and institutional buildings over 100,000 square feet.

Adding the CBECS and MECS stock and flow estimates derived for 2002 yielded the extrapolation base for this study.[19] Therefore, the total stock base was 38.650 billion square feet for all capital facilities; this is equivalent to 3.591 billion square meters. The total flow of new capital facility activity was 1.137 billion square feet, or 105.638 million square

---

[19]The *average* annual growth in capital square footage between 1994/5 and 1998/9 is used to represent construction activity in a typical year and is intended to minimize the variance in construction activity from year to year.

meters. Details on how these estimates were employed are presented in the following discussion.

5.4.2 Generating National Impact Estimates by Stakeholder Group, by Life-Cycle Phase

After cost estimates were developed by stakeholder group for each life-cycle phase, the estimates were divided by the total square footage stock (for the O&M phase) and flow (for the planning, engineering, and design and construction phases) provided by respondents. Therefore, each cost category had a cost per square foot estimate for each stakeholder group, by life-cycle phase. The cost per square foot estimate was then applied to the total national stock and flow estimates from CBECS and MECS.

Capital facilities development is an ongoing process. Therefore, the same national flow estimates were used for the planning, engineering, and design and construction phases for all stakeholders. However, the capital facilities stock estimates were applied only to owners' and operators' cost estimates for the O&M life cycle. Table 5-5 presents the extrapolation base for each stakeholder for each life-cycle phase.

Table 5-5. Extrapolation Base Data by Stakeholder, by Life-Cycle Phase

Capital Facilities Measure by Affected Stakeholder Group	Planning, Engineering and Design Phase: 1,137 million ft^2	Construction Phase: 1,137 million ft^2	Operations and Maintenance Phase: 38,650 million ft^2
Architects and Engineers	✓	✓	
General Contractors	✓	✓	
Specialty Fabricators and Suppliers	✓	✓	
Owners and Operators	✓	✓	✓

Sources: RTI estimates based on EIA, 1997; EIA, 1998; EIA, 2001b; EIA, 2002.

An illustration is useful to explain how cost estimates were extrapolated to national impact levels. Suppose that the average costs for all capital facilities owners and operators in the United States are estimated to be

- $0.05 per square foot for the planning, engineering, and design phase,
- $0.03 per square foot for the construction phase, and
- $0.20 per square foot for the O&M phase.

These cost estimates would then be applied to the extrapolation bases for each life-cycle phase for owners and operators to generate national impacts. These calculations would be

- $0.05 x 1,137 million square feet for the planning, engineering, and design phase, or $56 million;

- $0.03 x 1,137 million square feet for the construction phase, or $34 million; and

- $0.20 x 38,650 million square feet for the O&M phase, or $7,730 million.

The total, hypothetical inadequate interoperability cost estimate for owners and operators would be the sum of these three figures, or $7,821 million. The actual cost estimates presented in Chapter 6 provide greater detail than this example.

6 Estimated Costs of Inadequate Interoperability

The final two chapters of this report present quantitative and qualitative interoperability cost findings based on interviews and surveys conducted by RTI and LMI. Annual interoperability cost estimates presented in this section are calculated for 2002. These cost estimates primarily reflect inefficient activities and systems associated with managing and exchanging electronic and paper-based data. Chapter 7 presents additional qualitative findings about the interconnectedness of these costs between stakeholder groups and information technology adoption and usage, and discusses the issues and trends that impact the direction of electronic information exchange, management, and interoperability in the capital facilities supply chain.

Based on the methodology presented in Chapters 4 and 5, $15.8 billion in interoperability costs were quantified for the U.S. capital facilities supply chain in 2002 (see Table 6-1). This annual cost estimate corresponds to between 0.86 and 1.24 percent of annual receipts for architects and engineers, general contractors, and specialty fabricators and suppliers (see Table 6-2). When compared to the annual value of capital facilities construction put in place for 2002, owners and operators' total estimated costs are approximately 2.84 percent. It seems $15.8 billion is likely to be a conservative estimate because it does not include such cost categories as opportunity costs and decommissioning costs. Also, costs were not quantifiable for all the inefficiency cost components discussed in the preceding two chapters. Where costs were not able to be captured adequately, dashed lines are placed in this chapter's tables.

Table 6-1. Costs of Inadequate Interoperability by Stakeholder Group, by Life-Cycle Phase (in $Millions)

Stakeholder Group	Planning, Engineering, and Design Phase	Construction Phase	Operations and Maintenance Phase	Total
Architects and Engineers	1,007.2	147.0	15.7	1,169.8
General Contractors	485.9	1,265.3	50.4	1,801.6
Specialty Fabricators and Suppliers	442.4	1,762.2	—	2,204.6
Owners and Operators	722.8	898.0	9,027.2	10,648.0
Total	**2,658.3**	**4,072.4**	**9,093.3**	**15,824.0**

Source: RTI estimates.

Table 6-2. Costs of Inadequate Interoperability Compared to 1997 Establishment Revenue (A&E, GC, SF) and 2002 Value of Construction Set in Place (OO) (in $Millions)

Architects and Engineers[a]		
NAICS 54131	Architectural services	16,988.4
NAICS 54133	Engineering services	88,180.7
	Subtotal	105,169.0
Interoperability cost estimate ($)		1,169.8
Interoperability cost estimate (%)		1.11%
General Contractors[a]		
NAICS 2333	Nonresidential building construction	209,269.2
Interoperability cost estimate ($)		1,801.6
Interoperability cost estimate (%)		0.86%
Specialty Fabricators and Suppliers[a]		
NAICS 2351	Plumbing, heating, and air conditioning contractors	88,427.4
NAICS 2353	Electrical contractors	64,915.1
NAICS 23591	Structural Steel Erection contractors	8,152.7
NAICS 23592	Glass and Glazing contractors	4,045.5
NAICS 23594	Wrecking & Demolition contractors	2,304.0
NAICS 23595	Building equipment & other machinery installation contractors	9,342.9
	Subtotal	177,187.7
Interoperability cost estimate ($)		2,204.6
Interoperability cost estimate (%)		1.24%
Owners and Operators[b]		
Annual value of construction put in place, 2002		374,118.0
Interoperability cost estimate ($)		10,648.0
Interoperability cost estimate (%)		2.84%

[a]U.S. Census Bureau. 2004a. "1997 Economic Census: Summary Statistics for United States 1997 (NAICS Basis). http://www.census.gov/epcd/ec97/us/US000.HTM. As obtained on April 1, 2004.

[b]U.S. Census Bureau. 2004b. "Annual Value of Construction Set In Place." As released on April 1, 2004 at http://www.census.gov/const/C30/Total.pdf.

In fact, the majority of estimated costs were borne by owners and operators. The operations and maintenance phase has more cost associated with it than other life-cycle phases as information management and accessibility hurdles hamper efficient facilities operation. Owners and operators bore approximately $10.6 billion, or about two-thirds of the total estimated costs in 2002. Architects and engineers had the lowest interoperability costs at $1.2 billion. General contractors and specialty fabricators and suppliers bore the balance of costs at $1.8 billion and $2.2 billion, respectively.

These cost estimates were developed based on a year of interviews, focus groups, and an Internet survey in which 70 organizations participated. Average cost estimates per square foot were calculated by stakeholder group, life-cycle phase, and activity category. These per-unit impacts were then weighted by construction activity or capital facility stock to develop national impact estimates for the capital facility industry. Total new construction activity for 2002 was estimated to be approximately 1.1 billion square feet (106 million square meters). The total square footage set in place was estimated to be nearly 39 billion (3.6 billion square meters) (EIA, 1997; EIA, 1998; EIA, 2001b; EIA, 2002).

It is important to note that in some instances the data collected were not sufficient to quantify some cost categories that were investigated as part of the interviews and surveys.[20] This does not imply that certain types or categories of costs do not exist for a given stakeholder or during a given life-cycle phase. It simply means that the data collection and survey effort did not capture enough data to reliably quantify costs. In many instances, respondents indicated they had interoperability costs in an identified area, but were unwilling to "speculate" on the magnitude of the problem. Thus, the estimates presented in this report are likely to be conservative estimates of the interoperability inefficiency cost categories discussed in Chapters 4 and 5.

As shown in Table 6-3, most costs fall into the categories of mitigation and avoidance costs. Owners and operators engage primarily in mitigations costs and general contractors and specialty fabricators and suppliers engage primarily in avoidance costs. Quantified delay costs are primarily associated with owners and operators. However, all

---

[20]The term "quantify impacts" is used when discussing the results to emphasize that data could not be collected to estimate all interoperability costs. Thus, the cost impacts presented in this section represent a subset of the total interoperability costs.

Table 6-3. Costs of Inadequate Interoperability by Cost Category, by Stakeholder Group (in $Millions)

Cost Category	Architects and Engineers	General Contractors	Specialty Fabricators and Suppliers	Owners and Operators	Total
Avoidance Costs	485.3	1,095.4	1,908.4	3,120.0	**6,609.1**
Mitigation Costs	684.5	693.3	296.1	6,028.2	**7,702.0**
Delay Costs	—	13.0	—	1,499.8	**1,512.8**

Source: RTI estimates, totals may not sum correctly due to rounding.

stakeholder groups indicated that seamless exchange of electronic data would shorten design and construction time (even though they could not quantify the impact).

6.1 STUDY PARTICIPANTS

One hundred and five interviews representing 70 organizations contributed to the cost analysis of inadequate interoperability costs. These organizations provided the information, anecdotes, technical expertise, and data that support this analysis. Many organizations had multiple individuals participate in RTI's focus groups, telephone interviews, on-site visits, and Internet survey. Thus, the actual number of individuals providing information for the study far exceeded the number of organizations presented in Table 6-4.

Table 6-4. Study Participants by Stakeholder Group

Stakeholder Group	Number of Interviewees	Number of Organizations
Architects and Engineers	21	19
General Contractors	11	9
Specialty Fabricators and Suppliers	5	5
Owners and Operators	53	28
Software Vendors	5	2
Research Consortia	10	7
Total	**105**	**70**

Invitations to participate in this study were distributed by a variety of means. Announcements were made at industry conferences. In addition, several trade associations and industry consortia issued notifications to their members via their Web sites, newsletters, periodicals, and word of mouth. Several organizations also participated in early scoping interviews to help define the scope of this project. These organizations continued their participation through the entire effort.

A variety of avenues were available for stakeholders to participate in the study. Most organizations participated via the Internet survey that was housed at https://consint.rti.org. The Internet survey also permitted respondents to indicate whether they would like to continue/expand their participation. Willing organizations subsequently joined a substantial number of others that were involved in in-depth teleconferences, focus groups, and on-site visits. On-site visits were particularly helpful as they allowed the project team to spend on average one full business day speaking with representatives from all functional areas within an organization. The surveys that informed the analysis are included in Appendix B; the data collected from these instruments were supplemented by the more detailed information gathered during the on-site visits and telephone interviews.

Owners and operators were ultimately the most represented stakeholder group with 28 organizations participating. Architects and engineers were represented by 19 organizations. Fourteen general contractors and specialty fabricators and suppliers participated in the study. In addition, nine software vendors and research consortia contributed information concerning software applications, trends, and usage, and ongoing research and development efforts aiming to improve interoperability.

6.2 ARCHITECTS AND ENGINEERS

Most interoperability costs for architects and engineers understandably occur during the planning, engineering, and design phase; however, they also incur costs associated with the remaining life cycles. In large part, this is due to coordination with general contractors and owners and operators as construction progresses and facilities enter service. Quantified interoperability costs for architects and engineers are estimated to be $1.2 billion, with over 80 percent of these costs incurred during the design life-cycle phase. In addition, respondents indicated that seamless electronic information exchange and management could compress their schedules by as much as 10 percent. However, acceleration benefits, such as resource or asset depreciation costs, are not included in the cost estimate.

The key cost areas for architects and engineers include manual reentry, inefficient business process management, and RFI management costs. Table 6-5 breaks all quantified interoperability cost estimates out by cost category and component.

Manual reentry costs were $463 million during the design phase and $28 million during the construction phase. Stakeholders indicated that these costs have three sources. First, they result from translating and transferring electronic files between competing software packages. This occurs when different organizations are collaborating on a product and are using incompatible software. For example, an architect may be using AutoCAD for building design for a hospital, yet an engineering team in another organization may be using MicroStation to design HVAC and mechanical systems. Even when using translation software, staff in each organization frequently have to correct the geometry of the electronic design files.

A second source of manual reentry costs is the use of paper and electronic files in tandem. Each iteration of a design, or a component of a design, may be inputted from paper to CAD many times over, resulting in lost time for staff re-inputting data over and over again. One firm indicated that not all of their staff is trained on CAx systems, particularly at the senior level. Therefore, they rely on junior members of the project team to input paper design changes into electronic systems.

The third source of manual reentry problems stems from the receipt of paper design changes from external organizations. In these instances, staff must search through the paper files for changes and input them into the electronic files housed internally. Architects and engineers interviewed indicated that many of these changes come from general contractors and owner and operators who request that the electronic files be updated as facility installation progresses.

Manual reentry costs are interrelated with the three other activity cost categories:

- design and construction information verification costs,
- RFI management costs, and
- inefficient business process management costs.

Information verification costs totaled over $114 million during the planning, engineering, and design phase. As several electronic and paper designs for the same project circulate, staff members must ensure

Table 6-5. Costs of Inadequate Interoperability for Architects and Engineers

Life-Cycle Phase	Cost Category	Cost Component	Average Cost per Square Foot	Average Cost per Square Meter	Inadequate Interoperability Cost Estimate ($Thousands)
Planning, Engineering, and Design		Inefficient business process management costs	0.31	3.37	356,126
		Redundant CAx systems costs	0.0001	0.001	158
		Productivity losses and training costs for redundant CAx systems	0.04	0.45	47,947
		Redundant IT support staffing for CAx systems	0.0004	0.005	501
		Data translation costs	0.002	0.02	2,139
	Avoidance Costs	Interoperability research and development expenditures	0.02	0.21	22,234
		Manual reentry costs	0.41	4.38	462,734
	Mitigation Costs	Design and construction information verification costs	0.10	1.08	114,342
		Reworking design files costs	0.0009	0.009	968
		Avoidance costs	0.38	3.85	429,106
		Mitigation costs	0.51	5.47	578,044
	Subtotal	**Subtotal**	**0.89**	**9.32**	**1,007,150**
Construction	Avoidance Costs	Inefficient business process management costs	0.04	0.41	43,290
		Redundant CAx systems costs	0.00003	0.0003	28
		Productivity losses and training costs for redundant CAx systems	0.007	0.08	8,461
		Redundant IT support staffing for CAx systems	0.00008	0.0008	88
		Data translation costs	0.0003	0.004	378
		Interoperability research and development expenditures	0.003	0.04	3,924
		Manual reentry costs	0.024	0.26	27,750
	Mitigation Costs	Design and construction information verification costs	0.006	0.07	7,377
		RFI management costs	0.05	0.53	55,656
	Subtotal	Avoidance costs	0.05	0.49	56,169
		Mitigation costs	0.08	0.86	90,783
		Subtotal	**0.13**	**1.35**	**146,952**
Operations and Maintenance	Mitigation Costs	Post-construction redundant information transfer costs	**0.01**	**0.15**	**15,660**
Total Cost					**1,169,762**

Source: RTI estimates; totals may not sum correctly due to rounding.

that they are proceeding with the correct version of the design. This cost is incurred in large part to prevent further mitigation costs downstream. One firm cited that moving forward with the incorrect version of a file would become more costly as construction nears or progresses. The same firm considered this impact area to be a quality control issue. The cost, in their view, could be reduced but never fully eliminated even in the event of seamless electronic interoperability. Yet, the redundant activity is measurable and all interviewees believed that better interoperability would preempt most information verification costs.

During the construction phase, architects and engineers incur $55.7 million in RFI management costs, according to available data. These costs are primarily associated with general contractors requesting clarification on designs as construction progresses because information was not either adequately communicated or was not present in the documents provided. Respondent A&E firms spent an average of 200 hours each month managing RFIs.

Inefficient business process management costs account for almost $400 million, with 89 percent falling in the design phase. As information is exchanged between architects and engineers and other stakeholder groups, interoperability problems evolve from being concentrated solely within the realm of their functional area to business support services. These costs are best viewed first as supplemental coordination costs. Each instance of inadequate interoperability ripples into business support functions as staff request services to assist them in the resolution of inadequate interoperability problems. In addition, the lack of electronic communication between support services in different organizations requires organizations to build up support staffing levels to manage paper-based interactions.

The business management process costs represent the potential cost savings that could be gained from automating and integrating various process management systems in a firm or between stakeholders. There are many cases in which various systems and applications are designed to operate independently to solve specific requirements. The cost of interoperability is high because the business processes are not integrated. The key business processes where opportunities exist to reduce interoperability costs are

- project management,
- document management,

- information request processing,

- accounting, and

- cost estimation.

It was originally hypothesized that costs associated with the use and maintenance of redundant CAx systems would be a significant cost area. However, interviews and data from the surveys indicate that this is not the case. The low redundant CAx systems costs, in terms of lost productivity and systems support, is largely explained by the standardization of most architects and engineers on a small number of software systems. For example, most organizations reported using ArchiCAD, AutoCAD, or MicroStation, but not all three. Although firms appear to largely avoid these redundant CAx costs, they do incur costs when transferring costs between two or more competing software packages.

Costs also extend beyond life-cycle phases in which architects and engineers are typically directly involved. For example, $15.6 million in post-construction redundant information transfer costs were calculated between architects and engineers and owners and operators. Qualitative information from the interview portion of this study suggests that this may be an underestimate. Several organizations indicated that the process of reintegrating as-builts provided by general contractors, when required to do so, comprises a substantial amount of activity. In addition, several A&E firms stated that owners and operators frequently return to request further information well after facilities have entered service.

6.3 GENERAL CONTRACTORS

Based on the data available, general contractors incurred an estimated $1.8 billion in inadequate interoperability costs in 2002. Just over half of these costs, $1.1 billion, were inefficient business process management costs (see Table 6-6). Apart from interoperability research and development costs, no costs were able to be calculated for impact areas related to CAx systems, including redundant systems. This is not to suggest that these costs do not exist; they were unable to be adequately captured and disclosed over the course of this project. Quantified research and development expenditures suggest that such costs are in fact incurred.

Table 6-6. Costs of Inadequate Interoperability for General Contractors

Life-Cycle Phase	Cost Category	Cost Component	Average Cost per Square Foot	Average Cost per Square Meter	Inadequate Interoperability Cost Estimate ($Thousands)
Planning, Engineering, and Design	Avoidance Costs	Inefficient business process management costs	0.14	1.55	163,674
		Redundant CAx systems costs	—	—	—
		Productivity losses and training costs for redundant CAx systems	—	—	—
		Redundant IT support staffing for CAx systems	—	—	—
		Data translation costs	—	—	—
		Interoperability research and development expenditures	0.0006	0.006	630
	Mitigation Costs	Manual reentry costs	0.16	1.74	184,028
		Design and construction information verification costs	0.006	0.06	6,302
		RFI management costs	0.12	1.24	131,299
	Subtotal	Avoidance costs	0.14	1.55	164,304
		Mitigation costs	0.28	3.05	321,629
		Subtotal	**0.43**	**4.59**	**485,933**
Construction	Avoidance Costs	Inefficient business process management costs	0.82	8.78	927,487
		Redundant CAx systems costs	—	—	—
		Productivity losses and training costs for redundant CAx systems	—	—	—
		Redundant IT support staffing for CAx systems	—	—	—
		Data translation costs	—	—	—
		Interoperability research and development expenditures	0.003	0.03	3,571
	Mitigation Costs	Manual reentry costs	0.11	1.19	126,047
		Design and construction information verification costs	—	—	—
		RFI management costs	0.16	1.74	183,818
		Construction site rework costs	0.01	0.11	11,356
	Delay Costs	Idle employees costs	0.01	0.12	12,988
	Subtotal	Avoidance costs	0.82	8.78	931,059
		Mitigation costs	0.28	3.04	321,221
		Delay costs	0.01	0.12	12,988
		Subtotal	**1.11**	**11.94**	**1,265,268**
Operations and Maintenance	Mitigation Costs	Post construction redundant information transfer costs	**0.04**	**0.48**	**50,419**
Total Cost					**1,801,620**

Source: RTI estimates; totals may not sum correctly due to rounding.

General contractors as a whole had the lowest level of technology adoption of the interviewees. Although they maintain CAx systems, most of their work is performed using paper copies of design and engineering files. Respondents said this is because the work environment at the construction site precludes widespread usage of computing technologies. General contractors also believe that the return on investment for construction equipment is greater than the return on information technology, such as investing in a greater number of CAx stations.

The use of paper has several implications for business support functions. Inefficient business process management costs comprise $1.1 billion over both the planning, engineering, and design and construction phases. Interviewees stated that the majority of their work is done on paper and that paper is passed off to processing teams, including those for information requests, materials management and procurement, and inspection and certification. Respondents also reiterated the same coordination issues expressed by architects and engineers, in that collaborations require a significant amount of double entry into and among paper and electronic systems. These costs are captured as redundant labor costs. Respondents indicated that if these systems were fully electronic and interoperable, $1.1 billion could be trimmed from overhead budgets. The key areas in which opportunities exist to reduce interoperability costs are

- information request processing,
- document management,
- project management,
- procurement, and
- facility planning and scheduling.

Information request processing costs are complemented by information request management costs. The two cost areas were separated in this study because the former involves administrative employees and the latter, construction and project managers. According to respondents, general contractors make on average 350 RFIs per project during the design phase and 250 RFIs during the construction phase. In both instances, each RFI takes several hours to assemble. The average waiting time for a satisfactory response is 10 business days. RFI management costs are estimated to be $131 million during the first life-cycle phase and $184 million during the second.

Manual reentry costs comprised the third-largest cost category for general contractors. Costs were estimated to be $184 million during the design phase and $126 million during the construction phase, as construction managers and civil engineers move information between and among paper-based and electronic systems. One respondent cited a recently completed large-scale project that had three to four full-time employees on site performing these activities for the duration of the construction phase. Even after the construction phase ends, general contractors continue to incur costs as they transfer as-built information to architects and engineers and to owners and operators. These costs were estimated to be $50.4 million in 2002.

The consensus among general contractors is that seamless electronic information management and exchange would permit them to compress their schedules by an average of 7.5 percent (as with A&Es, more efficient resource utilization due to acceleration is not included in the cost estimates). General contractors cite incompatible computer systems, firewall limitations, and the reduction of paper-based information systems as key opportunities to achieve shorter completion schedules. The biggest challenge is the "lack of ability to have everyone on the same page with current information," according to one contractor. They also reported that facilities inspectors prefer to review paper information to ensure that buildings are up to code in addition to physical inspection.

6.4 SPECIALTY FABRICATORS AND SUPPLIERS

Inadequate interoperability costs for specialty fabricators and suppliers are estimated to be $2.2 billion for 2002, $400 million more than general contractors. Table 6-7 presents the impacts by cost category and component for this stakeholder group. Eighty percent of costs are incurred during the construction phase, primarily because of collaboration with general contractors to move the facility toward completion. The balance of costs are incurred during the design phase as specialty fabricators and suppliers coordinate with other stakeholder groups on design and engineering issues related to their technical area of expertise, such as structural steel associated with elevator systems.

As with general contractors and architects and engineers, interoperability problems extend beyond the technical focus of their work to business support services. These costs constitute the same internal and external coordination issues related to information exchange and management, both internally and externally, that were described for architects and

Table 6-7. Costs of Inadequate Interoperability for Specialty Fabricators and Suppliers

Life-Cycle Phase	Cost Category	Cost Component	Average Cost per Square Foot	Average Cost per Square Meter	Inadequate Interoperability Cost Estimate ($Thousands)
Planning, Engineering, and Design		Inefficient business process management costs	0.25	2.65	279,652
		Redundant CAx systems costs	0.0001	0.0007	70
		Productivity losses and training costs for redundant CAx systems	0.0002	0.002	230
		Redundant IT support staffing for CAx systems	—	0.0004	44
		Data translation costs	0.005	0.05	5,366
	Avoidance Costs	Interoperability research and development expenditures	0.0008	0.009	953
		Manual reentry costs	0.11	1.21	128,119
	Mitigation Costs	Design and construction information verification costs	0.02	0.19	20,019
		RFI management costs	0.007	0.08	7,944
	Subtotal	Avoidance costs	0.25	2.70	286,316
		Mitigation costs	0.14	1.48	156,081
		Subtotal	**0.39**	**4.18**	**442,397**
		Inefficient business process management costs	1.39	15.00	1,584,696
		Redundant CAx systems costs	0.0001	0.0007	70
		Productivity losses and training costs for redundant CAx systems	0.001	0.012	1,305
		Redundant IT support staffing for CAx systems	0.0002	0.0024	249
		Data translation costs	0.027	0.29	30,410
	Avoidance Costs	Interoperability research and development expenditures	0.005	0.05	5,402
		Manual reentry costs	0.10	1.10	115,726
		Design and construction information verification costs	0.01	0.15	16,015
	Mitigation Costs	RFI management costs	0.007	0.075	7,944
		Construction site rework costs	0.0003	0.004	374
	Delay Costs	Idle employees costs	—	—	—
	Subtotal	Avoidance costs	1.43	15.30	1,622,132
		Mitigation costs	0.12	1.33	140,059
		Delay costs	—	—	—
Construction		**Subtotal**	**1.55**	**16.63**	**1,762,190**
Total Cost					**2,204,588**

Source: RTI estimates; totals may not sum correctly due to rounding.

engineers and general contractors. The key areas in which specialty fabricators and suppliers believe that opportunities exist to reduce overhead labor charges with seamless electronic information and management include

- project management,
- facility planning and scheduling,
- information request processing,
- procurement, and
- product data management.

Specialty fabricators and suppliers also identified significant opportunities in the areas of facility planning and scheduling, facility simulation, materials management, and maintenance planning and management. Manual reentry costs totaled $245 million; these costs also exhibit the same characteristics as those for general contractors. However, specialty fabricators and suppliers differed from general contractors in two general ways. First, they exhibit quantifiable costs related to redundant CAx systems usage. These costs are related to

- redundant CAx systems maintenance,
- productivity losses on these systems, and
- redundant information support staffing.

Although these estimated costs are relatively small based on data collected, the information provided implies that specialty fabricators and suppliers were more likely to incur redundant systems costs to collaborate with teaming partners using different software tools. They also incurred more data translation costs than architects and engineers and general contractors.

Specialty fabricators and suppliers reported modest RFI management costs. This stakeholder group makes fewer RFIs, on average 75 per project, which lead to lower costs for RFI management. However, each RFI takes longer to process (15.5 hours) and the time between submission and receipt of a satisfactory response is longer at 12.5 days per RFI.

Fully electronic interoperable design and business support systems would, according to respondents, compress their schedules by 5 to 10 percent. They echoed general contractors in citing each party having the same information at the same time as an area of opportunity for the capital facilities supply chain.

6.5 OWNERS AND OPERATORS

Owners and operators have the largest interoperability costs of all the stakeholders: over $10.6 billion, or about 68 percent of the total $15.8 billion of inadequate interoperability costs calculated for the capital facilities supply chain (see Table 6-8; Appendix D presents measures of cost variability for this stakeholder group). This is because owners and operators carry the burden of ongoing interoperability costs during the operations and maintenance phase. They also bear costs during the planning, engineering, and design phase ($723 million) and during the construction phase ($898 million). These costs are associated with the same issues discussed for the preceding three stakeholder groups. Owners and operators have the added responsibility of ensuring that work proceeds according to their needs and specifications. They submit on average between 145 and 200 RFIs per project and wait approximately 6 days to receive a satisfactory response.

Eighty-five percent of owners and operators' interoperability costs are incurred during the operations and maintenance phase. Quantified costs were estimated at approximately $9 billion in 2002. Although inefficient business process management costs are significant during this phase ($1.6 billion), the costs related to facilities management and maintenance are even larger. When asked during an interview, one facilities manager responsible for 750,000 square feet stated that "too many ways to communicate [information] creates gaps and chasms." Underlying operation and maintenance phase costs are issues relating to the receipt, processing, and distribution of information, both from recently completed facilities and for existing facilities.

An inordinate amount of time is spent locating and verifying specific facility and project information from previous activities. For example, as-built drawings (from both construction and maintenance operations) are not routinely provided and the corresponding record drawings are not updated. Similarly, information on facility condition, repair parts status, or a project's contract or financial situation is difficult to locate and maintain.

Legacy data issues are a significant concern for owners and operators. Over the years, owners and operators receive and maintain information in a variety of different media: preferred electronic file formats, miscellaneous file formats, and paper information. This information does not always adequately reflect the true configuration of facilities either

Table 6-8. Costs of Inadequate Interoperability for Owners and Operators

Life-Cycle Phase	Cost Category	Cost Component	Average Cost per Square Foot	Average Cost per Square Meter	Inadequate Interoperability Cost Estimate ($Thousands)
Planning, Engineering, and Design	Avoidance Costs	Inefficient business process management costs	0.38	4.07	430,111
		Redundant CAx systems costs	—	—	—
		Productivity losses and training costs for redundant CAx systems	—	—	—
		Redundant IT support staffing for CAx systems	—	—	—
		Data translation costs	—	—	—
		Interoperability research and development expenditures	0.0039	0.042	4,422
	Mitigation Costs	Manual reentry costs	0.16	1.67	176,882
		Design and construction information verification costs	0.0056	0.061	6,415
		RFI management costs	0.092	0.99	104,966
	Subtotal	Avoidance costs	0.38	4.07	434,533
		Mitigation costs	0.25	2.73	288,263
		Subtotal	**0.64**	**6.80**	**722,796**
Construction	Avoidance Costs	Inefficient business process management costs	0.49	5.32	561,926
		Redundant CAx systems costs	—	—	—
		Productivity losses and training costs for redundant CAx systems	—	—	—
		Redundant IT support staffing for CAx systems	—	—	—
		Data translation costs	—	—	—
		Interoperability research and development expenditures	0.003	0.03	3,618
	Mitigation Costs	Manual reentry costs	0.15	1.59	167,975
		Design and construction information verification costs	0.0068	0.07	7,701
		RFI management costs	0.14	1.48	156,793
	Subtotal	Avoidance costs	0.50	5.32	565,544
		Mitigation costs	0.29	3.15	332,469
		Subtotal	**0.79**	**8.47**	**898,013**
Operations and Maintenance	Avoidance Costs	Inefficient business process management costs	0.04	0.46	1,638,915
		Redundant CAx systems costs	—	—	—
		Productivity losses and training costs for redundant CAx systems	—	—	—
		Redundant IT support staffing for CAx systems	—	—	—

(continued)

Table 6-8. Costs of Inadequate Interoperability for Owners and Operators (continued)

Life-Cycle Phase	Cost Category	Cost Component	Average Cost per Square Foot	Average Cost per Square Meter	Inadequate Interoperability Cost Estimate ($Thousands)
		Redundant facilities management systems costs[a]	0.01	0.13	456,064
		Productivity loss and training costs on redundant facility management systems	0.0003	0.0035	12,615
		Redundant facility management systems IT support staffing costs	0.0003	0.003	10,701
		Data translation costs	—	—	—
		Interoperability research and development expenditures	—	0.0005	1,659
	Mitigation Costs	O&M staff productivity loss	0.02	0.17	613,310
		O&M staff rework costs	0.0001	0.0011	3,952
		O&M information verification costs	0.12	1.33	4,790,159
	Delay Costs	Idled employees costs	0.04	0.42	1,499,839
	Subtotal	Avoidance costs	0.05	0.59	2,119,954
		Mitigation costs	0.14	1.51	5,407,420
		Delay costs	0.04	0.42	1,499,839
		Subtotal	**0.23**	**2.51**	**9,027,214**
Total Cost					**10,648,023**

[a] The variability of costs for redundant facilities management systems is not presented in Appendix D to prevent the disclosure of individual survey responses.

Source: RTI estimates; totals may not sum correctly due to rounding.

because as-built information was poorly communicated or because information was poorly maintained over the years. The net result is that owners and operators suffer significant efficiency losses each year.

Efficiency losses mostly impact facilities management and operations and maintenance staff. The single largest impact is on information verification and validation, or the time spent ensuring that the information accurately represents what is set in place. These costs are estimated to be $4.8 billion in labor charges. Once the information is found, operations and maintenance engineers spend time valued at $613 million transferring information into a format that staff members can read and use to perform their activities.

The owners and operators who participated in this analysis indicated that 71 percent of design and engineering information is in paper format, 21 percent is in preferred electronic formats, and 8 percent is in miscellaneous electronic formats. However, these distributions varied greatly and the large majority of respondents indicated that over 90 percent of this information is in paper format. Only a few respondents indicated that a significant share of their information was in electronic form.

Information delays lead to idled employees waiting for information in order to resolve a facilities maintenance issue. These costs were estimated to be $1.5 billion in 2002. Furthermore, if they proceeded with inadequate information, they frequently need to revisit maintenance problems to resolve them correctly. As a result, the cost of inadequate interoperability for operations and maintenance staff is $6.9 billion.

Facilities management also incurs costs related to redundant information technology systems. Owners and operators commonly purchase multiple software packages with the same functionality. This is especially the case for larger organizations whose facilities management staff is divided into several functional teams that make independent business decisions. Redundant facilities management systems costs were $456 million in 2002. According to interviewees, the productivity loss on redundant systems in terms of redundant labor charges for users ($12.6 million) and IT support ($10.7 million) is less than the maintenance cost. Interviewees indicated that facilities managers do not want to lose information inputted and stored in redundant systems, nor do they wish to incur the costs associated with transferring that information into their preferred management systems. Management may therefore opt to maintain these systems to safeguard their data, even if these systems are no longer used regularly. The redundant facilities maintenance costs also include those costs for software packages purchased but never used.

Inefficient business process management costs were $2.6 billion. The areas in which the greatest opportunities exist to reduce overhead labor charges with seamless electronic information management and exchange are

- document management,
- maintenance planning and management,
- information request processing,
- facility planning, and
- project management.

Because owners and operators bear significant costs in all life-cycle phases and the five business processes listed above were significant cost drivers across all life-cycle phases, there may be opportunities for barrier removal if some of these business processes result in significant cost burdens for the three other stakeholder groups. A review of Tables 6-5, 6-6, and 6-7 and the relevant sections of the text reveals that information request processing and project management are key cost drivers for each of the three other stakeholder groups. Addressing the sources of these inefficiencies across all stakeholder groups may be a target of opportunity for gains in efficiency that can be shared by all stakeholders.

7 Issues, Drivers, and Future Trends

This chapter augments the empirical results presented in Chapter 6 with a discussion of the issues and drivers that underlie the estimated cost impacts. In addition to identifying and quantifying costs, a further objective of the study was to collect insights concerning opportunities to improve connectivity across stakeholder groups and within organizations. Thus, interviews with participants included prospective discussions that focused on the disconnects within the capital facilities supply chain and the opportunities that exist to eliminate these inefficiencies.

Stakeholders emphasized the interconnectedness of the inadequate interoperability costs they incur. Owners and operators, in particular, were able to illustrate the challenges of information exchange and management due to their involvement in each phase of the facility life cycle. In summary, they view their interoperability costs during the operations and maintenance phase as a failure to manage activities upstream in the design and construction process. Poor communication and maintenance of as-built data, communications failures, inadequate standardization, and inadequate oversight during each life-cycle phase culminate in downstream costs. This can be seen in the quantification of substantial costs related to inefficient business process management and losses in productivity for operations and maintenance staff.

However, owners and operators were not the only ones to express such frustrations regarding the costs they bear. During interviews with the three other stakeholder groups, many of the same issues were discussed. They reported that interoperability costs do not simply result from a failure to take advantage of emerging technologies, but stem from a series of disconnects, both within and among organizations, that contribute to redundant costs.

7.1 STAKEHOLDERS' VIEWS ON THE CHALLENGES AND IMPEDIMENTS TO IMPROVED INTEROPERABILITY

Different stakeholders are involved in the multiple phases of the facility life cycle, and they typically have limited contractual incentive to communicate. For example, there is typically little interaction between architects and HVAC suppliers. Interviewed stakeholders said that opportunities for improvement are lost due to the fact that these parties rarely communicate about their related responsibilities. One issue is that there are minimal incentives for architects to give continuously-updated information to other players beyond what is necessary given liability concerns.

Another issue raised during stakeholder interviews was how to provide incentives to software vendors to integrate data and information standards into their systems. For example, CAD vendors may be reticent to suggest solutions due to market share issues and loss of competitive advantage. The dominant vendor in the CAD marketplace's system is used by approximately 70 percent of architects in the United States. There is little incentive to change; an inability to extract data from their files that can be used by other systems is not a pressing issue for them and may indeed be a way of preserving their market share.

During facility development, software users are reluctant to incur additional costs that may yield benefits down stream but lack immediately tangible benefits. As one project team noted, "We did not complete as-builts on paper, never mind electronic. We ran out of money and had to make some hard decisions. [After the facility was in operation a year], we needed to add an elevator to the basement. We ended up surveying that area of the building and performing the as-builts for that area."

Similarly, equipment providers want to control the procurement process and the interface between themselves and customers. Participating in a consortium designed to resolve interoperability problems might impact these companies' competitiveness. Some owner and operator stakeholders believed that some suppliers participated only to gather market intelligence, rather than to make progress.

Frequently, the next party in the supply chain can do their job better, faster, and cheaper with electronic information, but the firm who did the design work may not be compensated fully for the resources invested to provide such information. That the architects and engineers stakeholder group bears the lowest cost of interoperability of the four groups magnifies the lack of incentives to interoperate. There is no established practice that provides incentives to groups to coordinate for the ultimate good of an owner and operator, except the potential for repeat business, which is not always guaranteed due to the industry's competitive bidding practices.

Another major challenge is the organization of information required to make rapid business decisions. Electronic mail management across the integrated team members is one issue cited. For instance, a project member leaves an organization, and there is typically no way to manage their institutional knowledge or e-mail. Version and configuration control are frequently impacted by employee turnover.

7.2 STAKEHOLDERS' VIEWS ON CONNECTIVITY OPPORTUNITIES

Comments from stakeholders in the capital facilities industry generally were similar to interoperability concerns that have been voiced over the past few decades:

- Delivery models must motivate all project participants to optimize value from the end result, and motivations must be tied to financial gain. Owners and operators and other participants must identify project goals and metrics and experiment with contract alternatives that link participants' financial motivations with the project's goals.

- The industry must develop tools to integrate across multiple disciplines and must link motivations and optimization around project value. Participants need tools that allow them to share information on a real-time basis (Beck, 2001).

During the interviews, stakeholders highlighted the following connectivity opportunities for improvement:

- Increased connectivity between CAx, facilities management, and information databases, such as electronic document management (EDM) and enterprise resource planning (ERP) systems, is needed. This would integrate the graphical and database systems for more effective use in decision making. One architect stated that "typically, our projects do not involve the electronic exchange/conversion of design data with operations and maintenance systems." Facilities managers cited this as the key opportunity for more efficient and effective maintenance of capital facilities assets.

- Facilities management systems should interoperate with building control systems to enhance operations and maintenance staff productivity. As one stakeholder indicated, "For every dollar you would save connecting design and construction, you could save several times that connecting O&M controls to the original CAD and engineering designs. Downstream is much more important than connecting design software to construction [given that facilities will be in operation for many years]. The design needs to have better connectivity to real world data."

- Greater use of neutral format standards could help interoperability. According to one project manager for a large owner operator, "The classic problem is the challenge of converting and the improper conversion of files from [one system's format] to [another system's] format and the associated errors in the final project." Inadequate interoperability is compounded by merging paper design versions with poorly converted electronic versions, "resulting in partial rework to have the converted files match hard copy submissions previously reviewed and approved."

- Management of correspondence files would help organize and manage the large volume of communication information received during project planning and execution.

- Universal acceptance of electronic signatures via regulatory bodies and contract participants would help eliminate the requirement for a paper-based purchasing system.

- Fundamental roadmap agreements between stakeholders are needed. A fundamental challenge set forth by the owners, such as, "On this date we will not buy anything unless it is compliant by this date and we will tell you how to make it compliant" could force the vendors to adjust their technology to meet the needs of the owners.

- Linkage of schedule and cost data would improve interoperability between different systems used in an enterprise, such as those for human resources (HR), finance, project management, accounting, etc.

- Building engineers need access via handheld devices so that they could connect work order systems to handheld devices.

- Radio frequency identification (RFID) tags for materials inventory management, and their integration with on-site software tools, would streamline business processes at construction sites and reduce delay costs.

- Building information models that connect CAx files to facilities management and building control systems would result in more effective management of facilities across all life-cycle phases.

- With a focus/concern on homeland security, physical and cyber security are current concerns that owners and operators (in addition to others who support them) will be facing. This is reflected in both the public and private sectors.

- Increasing the use of GIS-capable tools could increase efficiency.

7.2.1 On-Line Collaboration Tools

Stakeholders indicated that a recent trend in project management for architects and engineers and general contractors is the use of on-line project management and collaboration software. Collaboration software applications gained momentum in the late 1990s. A large number of applications can be considered under the umbrella term "project collaboration software." They range from simple Internet-based CAD drawing viewers designed to share drafts and drawings with project team members to entire online construction project management solutions provided through Application Service Providers (ASPs); Engineering, Procurement, and Construction (EPC) suppliers; and Design-Build business process software.

The adoption of online collaboration tools has accelerated in recent years because of mandates by owners and operators in the bidding process. Internet sites can be project specific, and from these sites, project participants can access the most current documents and changes. A key benefit is improved efficiency because employees have access to project information from remote offices. That access permits time savings, cost savings, and accountability.

Constructware is an example of an online project collaboration and management tool. The system offers online collaboration services for Web-based project management. Constructware has developed an XML schema that allows a project member to send files to a centralized database, where they can be viewed by the recipient and other members of a project team. A proprietary rules and routing capability allows users to specify how documents flow among the other team members. A CFMA report on construction industry collaboration software said that 35 percent of construction firms with annual revenues above $100 million used some kind of collaboration software, and that over half of the firms in this group used Constructware (McGraw Hill Construction, 2002).

The use of online collaboration tools is generally initiated by owners and operators who want to add accountability to the facility development process. Owners and operators are motivated to adopt such tools because they incur the most risk due to limited knowledge of how contracting service firms drive up project costs. Industry experts estimate that around 85 percent of owners and operators in the capital facilities industry are largely uninformed on issues related to project costs. They rely on the integrity of architects, engineers, and general contractors to keep projects within specified budget constraints. Excessive change orders (COs), and requests for information (RFIs) between general contractors, suppliers, and design teams can significantly affect the project's total cost.

Larger construction projects often have a large number of contractors all sharing information and designs. Finding documents can become a huge delay cost in itself. Collaboration tools allow instant updating or notification of new modifications to a file and the file location so that other project members can work on the same file. This organizational approach ensures that project team members can operate efficiently. The collaboration tool offers additional benefit by minimizing the number of project managers needed to oversee the work being performed by contractors.

A report by PriceWaterhouseCoopers cited the benefits of adopting online collaboration tools (Wesek, Cottrez, and Lander, 2000):

- Improved project progress communication.
- Reduced response time for RFIs, COs, and design clarification.
- Shortened time to completion.
- Increased ownership of the construction process by owners and operators and accountability for contractors.
- Improved record keeping and documentation.

7.3 STAKEHOLDERS' DRIVERS FOR IMPROVING INTEROPERABILITY

High-priority objectives identified by the stakeholders include making the procurement, project management, construction, and financial systems communicate with each other. Stakeholders would also like to improve CAD interoperability. It is widely perceived that CAD can be a more effective tool for the industry and provide more intelligent information for builders, operators, and managers. It can be enhanced so users can use the same data and program in the design and construction phases. This will reduce data reentry. Thus, stakeholders view the establishment of industry-wide data standards as a key objective.

However, multiple stakeholders expressed frustration about the lack of an obvious solution to the problem of interoperability within the capital facilities industry when asked the question of "who will take the lead in solving the problem?" Respondents said it was unclear and that it would depend on when and where financial incentives materialized.

Most stakeholders believed that owners and operators may be the key to solving interoperability issues, because they set and drive business and system requirements and have the strongest incentive because of the bulk of the burden of lack of interoperability. The majority of respondents felt that owners and operators will have to champion the cause, and that they should be charged with leading the effort to develop and implement interoperability solutions. Top management commitment is required to identify the players and facilitate agreement among them. This was done successfully in the semiconductor industry to establish interoperability standards. However, most owners and operators interviewed currently do not see the financial incentives for "stepping up to the plate" to improve interoperability unless it is done on a smaller scale or trial basis.

One approach would be for a consortium of owners and operators to lead the effort in well-defined geographical centers where competition for service providers (e.g., contractors, A&Es) would not be affected by the push for interoperable systems or electronic as-builts. As part of this, the owners may have to mandate interoperable systems (i.e., the data must be able to move easily and reliably from one application to another). Several groups are currently working on technical solutions to support such activities, including IAI and FIATECH, as they try to develop standards for data exchange and enhanced interoperability. However, an impediment to this approach is resistance from the large firms if they perceive they lose a competitive advantage by making their competitors more interoperable.

Few respondents supported an approach where the federal government would provide the mandate that owners adopt interoperability solutions. Internationally, for example, the Singapore government mandates that contractors use applications that are capable of sharing data with subcontractors that work for them. However, in the United States it was felt that the capital facilities industry is too fragmented and diverse to mandate behavior and that market forces should shape the industry's response.

Most respondents thought that NIST should work with the dominant firms in the capital construction business (owners, suppliers, contractors, software vendors) and trade associations and research organizations to promote interoperability through activities such as developing open standards. For instance, FIATECH, CII, and IAI are working in this area, but stakeholders felt that improved coordination among these organizations and philosophical agreement on how best to resolve the issues would be helpful. NIST could help coordinate activities and provide high visibility for their efforts.

One property manager reported that so many organizations produce similar operational data (such as the Institute of Real Estate Management and the Building Owners and Managers Association [BOMA] International) that it is difficult to know whose information to use. This manager recommended that a single national organization be tasked with leading the effort to organize and validate the information needed to address interoperability problems.

Finally, all stakeholders thought that business process issues should drive system requirements at the enterprise level and that private-sector needs should drive the eventual adoption of interoperability solutions.

References

ASTM International. 2002. "Standard Classification for Building Elements and Related Sitework – UNIFORMAT II." West Conshohocken, PA: ASTM International.

Bayramoglu, S. 2001. "Partnering in Construction: Improvement through Integration and Collaboration." *Leadership and Management in Engineering*, July 2001, p. 39-43.

Beck, P. 2001. "The AEC Dilemma—Exploring the Barriers to Change." *Leadership and Management in Engineering*, April, p. 31-36.

Blackman, Maury. January 2001. "AEC Community Seeks Common Language." *The American City & County*. 166(1):12. Pittsfield: January 2001.

Bowen, Brian, Robert Charette, and Harold Marshall. 1992. *UNIFORMAT II—A Recommended Classification for Building Elements and Related Sitework*. NIST Special Publication 841. Gaithersburg, MD: National Institute for Standards and Technology.

Bureau of Economic Analysis (BEA). 2002. *Survey of Current Business*. Washington, DC: U.S. Bureau of Economic Analysis.

Chapman, Robert. 2000. Benefits and Costs of Research: A Case Study of Construction Systems Integration and Automation Technologies in Industrial Facilities. NISTIR 6501. Gaithersburg, Maryland: National Institute for Standards and Technology.

Chapman, Robert. 2001. Benefits and Costs of Research: A Case Study of Construction Systems Integration and Automation Technologies in Commercial Buildings. NISTIR 6763. Gaithersburg, Maryland: National Institute for Standards and Technology.

Chinowsky, P.S. 2001. "Construction Management Practices Are Slowly Changing." *Leadership and Management in Engineering*, April, p. 17-22.

Cleland, David I. 1999. *Project Management: Strategic Design and Implementation.* 3rd Ed. New York: McGraw Hill.

Construction Financial Management Association (CFMA). 2002. *CFMA's 2002 Information Technology Survey For the Construction Industry.* CFMA: Princeton, N.J.

Cotts, David G. 1998. *The Facility Management Handbook.* 2nd Ed. New York: AMACOM.

Defense Systems Management College (DSMC). 2001. "The Program Manager's Notebook: Systems Engineering Advanced Program Management Course." Fort Belvoir, VA: Defense Acquisition University.

Eastman, Chuck. 2001. *Overview of CIS/2.* As obtained on August 29, 2003, from <http://www.coa.gatech.edu/~aisc/document/newCIS-2overview2.pdf>.

Energy Information Administration (EIA). 1997. "Manufacturing Consumption of Energy 1994." Washington, DC: U.S. Department of Energy.

Energy Information Administration (EIA). 1998. "Commercial Buildings Energy Consumption Survey 1995." Washington, DC: Energy Information Administration.

Energy Information Administration (EIA). 2001a. "Annual Electric Generator Report—Utility." As obtained on April 6, 2003, from <http://www.eia.doe.gov/cneaf/electricity/ipp/html1/ippv1te1p1.html>.

Energy Information Administration (EIA). 2001b. "The 1998 Manufacturing Energy Consumption Survey." Washington, DC: Energy Information Administration.

Energy Information Administration (EIA). 2001c. "Petroleum Supply Annual 2001." Washington, DC: Energy Information Administration.

Energy Information Administration (EIA). 2002. "The 1999 Commercial Building Energy Consumption Survey." Washington, DC: Energy Information Administration.

FIATECH. 2002. "Capital Projects Technology Roadmapping Initiative." As obtained on January 28, 2003, from <http://www.fiatech.org/projects/cptri.htm>.

Fogel, Robert W. March 1979. "Notes on the Social Saving Controversy." *The Journal of Economic History* 39(1):1-54.

Gale Group. 2001c. *Encyclopedia of American Industries, 3rd ed.* "General Contractors—Nonresidential Buildings, Other Than Industrial Buildings and Warehouses. Reproduced in Business and Company Resource Center. Farmington Hills, MI: Gale Group. As obtained on May 20, 2003, from <http://www.galenet.com/servlet/BCRC>

Gale Group. 2001a. *Encyclopedia of American Industries, 3rd ed.* "Architectural Service." Reproduced in Business and Company Resource Center. Farmington Hills, MI: Gale Group. As obtained on May 20, 2003, from <http://www.galenet.com/servlet/BCRC>.

Gale Group. 2001b. *Encyclopedia of American Industries, 3rd ed.* "General Contractors—Industrial Buildings and Warehouses." Reproduced in Business and Company Resource Center. Farmington Hills, MI: Gale Group. As obtained on May 20, 2003, from <http://www.galenet.com/ servlet/BCRC>.

Gale Group. 2001d. *Encyclopedia of Global Industries, 3rd ed.* "Real Estate Investment Trusts." Reproduced in Business and Company Resource Center. Farmington Hills, MI: Gale Group. As obtained on May 20, 2003, from <http://www.galenet.com/ servlet/BCRC>

Gallaher, Michael, Alan O'Connor, and Thomas Phelps. 2002. Economic Impact Assessment of the Standard for the Exchange of Product Model Data (STEP) in Transportation Equipment Industries. Prepared for the National Institute for Standards and Technology, December 2002.

Hudson, W. Ronald, Ralph Haas, and Waheed Uddin. 1997. *Infrastructure Management.* New York: McGraw-Hill.

International Alliance for Interoperability-North America (IAI-NA). 2003. "aecXML: Mission." As obtained on August 8, 2003, from <http://www.iai-na.org>.

International Organization of Standards (ISO). 2003. "Standard Generalized Markup Language (SGML) ISO 8879:1986." As obtained on August 8, 2003, from <http://www.iso.ch/iso/en/CatalogueDetailPage.CatalogueDetail? CSNUMBER=16387&ICS1=35&ICS2=240&ICS3=30>.

Luiten, G.T., and F.P. Tolman. 1997. "Automating Communication in Civil Engineering." *Journal of Construction Engineering and Management,* June, p. 113-120.

McGraw Hill Construction. 2002. *Constructware Ranked #1 in CFMA Survey of Construction Industry Collaboration Software.* As obtained on October 10, 2003, at <www.construction.com>.

Martin, Sheila, and Smita Brunnermeier. 1999. Interoperability Cost Analysis of the U.S. Automotive Supply Chain. Prepared for the National Institute for Standards and Technology, March 1999.

National Institute of Standards and Technology (NIST). 2002. "CIS/2 and VRML Research at NIST." As obtained on August 29, 2003, from <http://cic.nist.gov/vrml/cis2.html>.

National Research Council (NRC). 1990. *Pay Now or Pay Later: Controlling Cost of Ownership from Design Throughout the Service Life of Public Buildings.* Washington, DC: National Academy Press.

National Research Council (NRC). 1998. *Stewardship of Federal Facilities: A Proactive Strategy for Managing the Nation's Public Assets.* Washington, DC: National Academy Press.

O'Reilly and Associates, Inc. 2003. "XML: Frequently Asked Questions (FAQs)." As obtained on August 8, 2003, from <http://www.xml.com>.

Roe, Andrew, and Peter Reina. August 13, 2001. "Learning to Share is Tougher than Anyone Anticipated: Industry fragmentation impedes automation." *Engineering News Record (ENR).* 247(7). McGraw Hill: New York.

Sullivan, William G., Elin M. Wicks, and James T. Luxhoj. 2003. *Engineering Economy.* 12th Ed. Upper Saddle River, New Jersey: Prentice Hall.

Tulacz, Gary J., and Debra K. Rubin. 2002. "Owners Turn Up the Heat." *McGraw-Hill Construction/Engineering News-Resource.* As obtained on May 20, 2003, from <http://enr.construction.com>.

Tulacz, Gary J., and Mary B. Powers. 2003. "Broad Downturn Touches Most." *McGraw-Hill Construction/Engineering News-Resource.* As obtained on May 20, 2003, from <http://enr.construction.com>.

Tulacz, Gary J., Debra K. Rubin, and Tom Armistead. 2003. "Market Jitters Worry Design Firms." *McGraw-Hill Construction/ Engineering News-Resource.* As obtained on May 20, 2003, from <http://enr.construction.com>.

U.S. Census Bureau (Census). 2000a. 1997 Economic Census. Construction Subject Series. EC97C23S-IS. Table 1. As obtained on June 5, 2003, from <http://www.census.gov/ prod/ec97/97c23-is.pdf>.

U.S. Census Bureau (Census). 2000b. 1997 Economic Census. Professional, Scientific, and Technical Services Geographical Area Series. EC97S54A-US. Table 1. As obtained on June 5, 2003, from <http://www.census.gov/ prod/ec97/97s54-us.pdf>.

U.S. Census Bureau (Census). 2000c. 1997 Economic Census. Real Estate and Rental and Leasing Geographical Area Series. EC97F53A-US(RV). Table 1. As obtained on June 5, 2003, from <http://www.census.gov/prod/ ec97/97f53-us.pdf>.

U.S. Census Bureau. 2004a. "1997 Economic Census: Summary Statistics for United States 1997 (NAICS Basis). http://www.census.gov/epcd/ec97/us/US000.HTM. As obtained on April 1, 2004.

U.S. Census Bureau. 2004b. "Annual Value of Construction Set In Place." As released on April 1, 2004 at http://www.census.gov/const/C30/Total.pdf.

U.S. Department of Labor, Bureau of Labor Statistics (BLS). 2003a. *Career Guide to Industries, 2002-03 Edition, Construction.* As obtained on May 20, 2003, from <http://www.bls.gov/oco/cg/cgs003.htm>.

U.S. Department of Labor, Bureau of Labor Statistics (BLS). 2003b. *Occupational Outlook Handbook, 2002-03 Edition, Architects, Except Landscape and Naval.* As obtained on May 20, 2003, from <http://www.bls.gov/oco/ ocos038.htm>.

U.S. Department of Labor, Bureau of Labor Statistics (BLS). 2003c. *Occupational Outlook Handbook, 2002-03 Edition, Construction Managers.* As obtained on May 20, 2003, from <http://www.bls.gov/oco/ocos005.htm>.

U.S. Department of Labor, Bureau of Labor Statistics (BLS). 2003d. *Occupational Outlook Handbook, 2002-03 Edition, Elevator Installers and Repairers.* As obtained on May 20, 2003, <http://www.bls.gov/oco/ocos189.htm>.

U.S. Department of Labor, Bureau of Labor Statistics (BLS). 2003e. *Occupational Outlook Handbook, 2002-03 Edition, Heating, Air-Conditioning, and Refrigeration Mechanics and Installers.* As obtained on May 20, 2003, from <http://www.bls.gov/oco/ocos192.htm>

U.S. Department of Labor, Bureau of Labor Statistics (BLS). 2003f. *Occupational Outlook Handbook, 2002-03 Edition, Structural and Reinforcing Iron Metal Workers.* As obtained on May 20, 2003, from <http://www.bls.gov/oco/ ocos215.htm>

U.S. Bureau of Labor Statistics (BLS). 2004. "2002 Occupational Employment Statistics (OES)." As obtained on March 12, 2004, from <http://www.bls.gov/oes/home.htm>.

U.S. General Services Administration (GSA). 2003. "Public Buildings." As obtained on June 5, 2003. <http://www.gsa.gov/Portal/browse/channel.jsp?channelId=-13820&channelPage=/channel/default.jsp>.

Wesek, Jacqueline, Vincent Cottrez, and Philip Lander. 2000. *A Benefits Analysis of Online Project Collaboration Tools within the Architecture, Engineering and Construction Industry.* Price Waterhouse Coopers.

Appendix A:
Description of the Typical
Business Process

There are six major phases in the capital facilities lifecycle. Though for the purposes of this report, they were condensed into four. All six are presented here for illustrative purposes. As presented in Figure A-1, they are:

1. Planning and Programming
2. Engineering and Design
3. Construction
4. Commissioning
5. Operation and Maintenance (O & M)
6. Disposal

The following figures depict the activities and organizations within each phase. Throughout the 20+ year facility lifecycle, there are many organizations involved in executing these activities, such as these major organizations:

- Customer, end-user
- Facility manager
- Designer, architectural, and engineering firm
- Contractor
- Specialty subcontractors
- Equipment and material suppliers

Additionally, there are many organizations and entities that require access to information about the facilities: insurance companies, utility companies, banks, local and state jurisdictions, adjacent property owners, building tenants, just to name a few.

Figures A-2 through A-9 provide greater detail about the typical business processes within each of the phases. Data exchange between the organizations and entities identified above occurs according to different requirements of each organization throughout the life cycle of the facility. For instance, costing and budget data and information related to the programming, design, construction, and operations and maintenance of the facility must be exchanged with the managers of the facility and the customers/end-users. Deriving these costs involves knowing and understanding the physical characteristics of the facility (e.g., intended use of the facility, total size and scope, initial and final design and layout, actual construction results [the "as-built" condition], and the operations and maintenance requirements). These interdependent physical characteristics drive the cost and budget data and information. Having high quality, interoperable systems that capture and maintain this data and information tends to improve effectiveness and efficiencies in managing facilities and ultimately reduces costs.

Figure A-1. Phases of the Capital Facilities Life Cycle

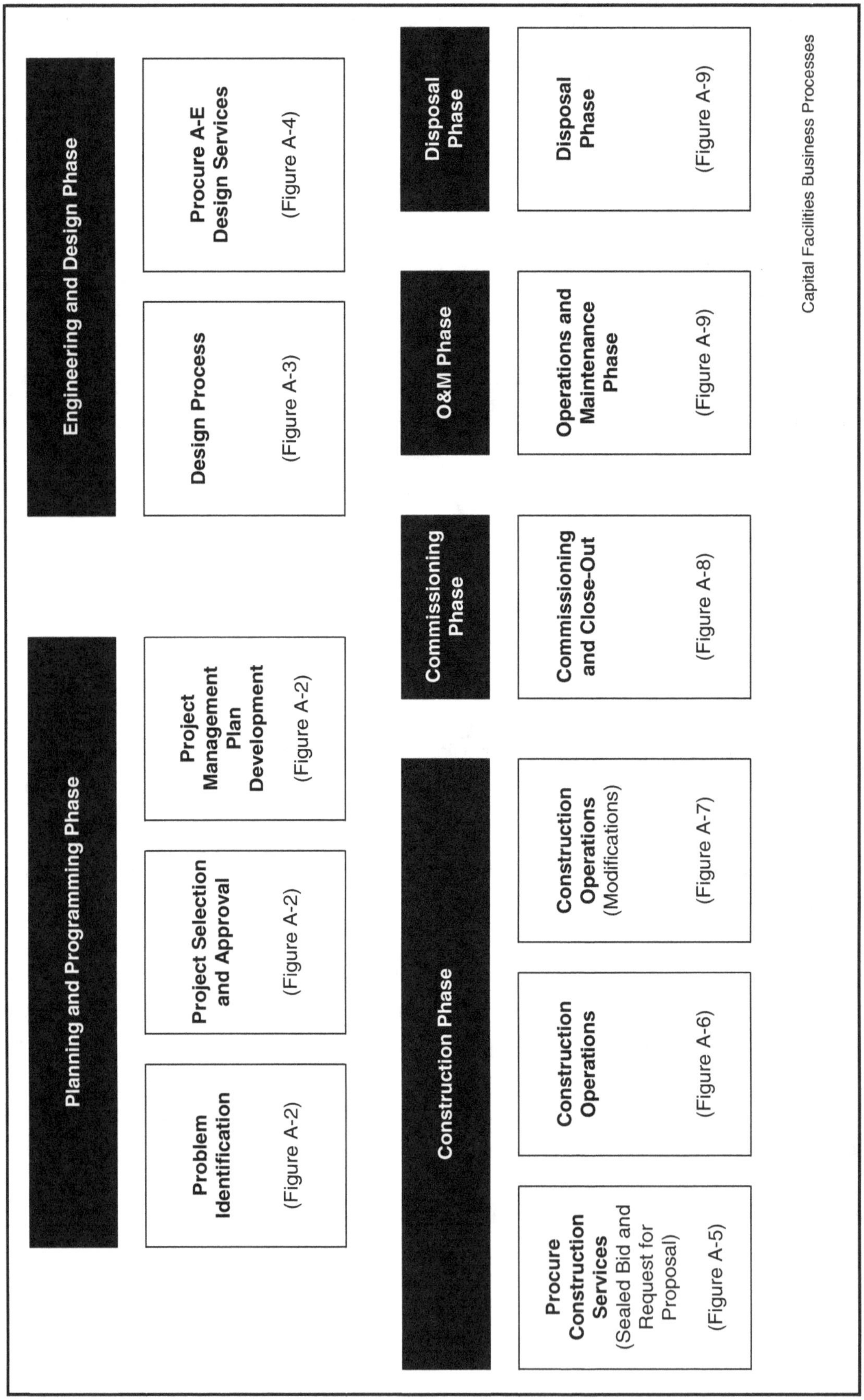

Planning and Programming Phase

Problem Identification	Project Selection and Approval	Project Management Plan Development
(Figure A-2)	(Figure A-2)	(Figure A-2)

Engineering and Design Phase

Design Process	Procure A-E Design Services
(Figure A-3)	(Figure A-4)

Construction Phase

Procure Construction Services (Sealed Bid and Request for Proposal)	Construction Operations	Construction Operations (Modifications)
(Figure A-5)	(Figure A-6)	(Figure A-7)

Commissioning Phase

Commissioning and Close-Out
(Figure A-8)

O&M Phase

Operations and Maintenance Phase
(Figure A-9)

Disposal Phase

Disposal Phase
(Figure A-9)

Capital Facilities Business Processes

Figure A-2. Typical Business Process of Capital Facilities Project—Planning and Programming Phase

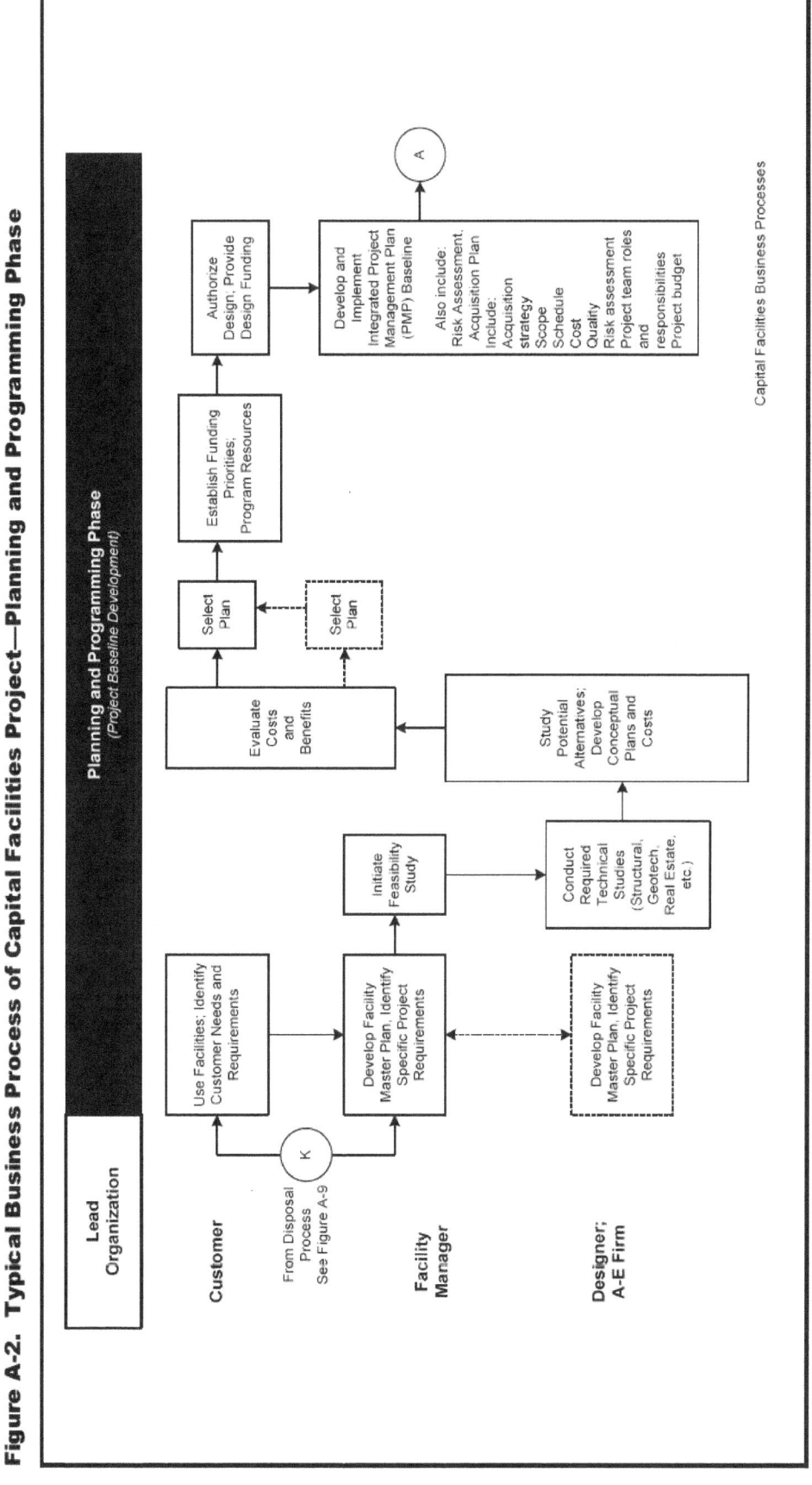

Figure A-3. Typical Business Process of Capital Facilities Project—Engineering and Design Phase: Design Process for Capital Facilities Projects

Engineering and Design Phase
(Design Process for Capital Facilities Projects)

Capital Facilities Business Processes

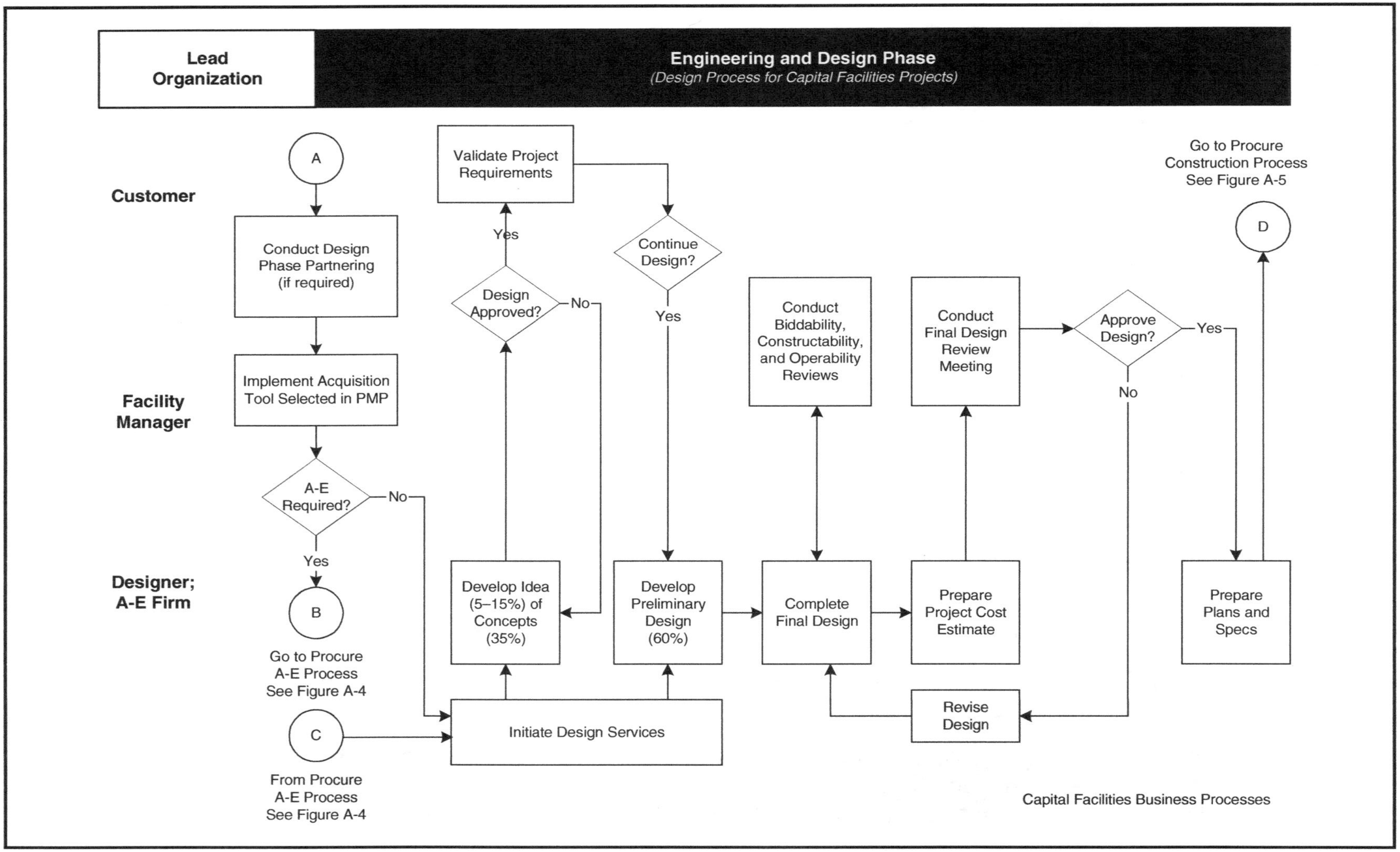

Figure A-4. Typical Business Process of Capital Facilities Project—Engineering and Design Phase: Procure A-E Design Services

Engineering and Design Phase
(Procure A-E Design Services)

Lead Organization

Customer

B

Establish Evaluation Criteria; Prepare Statement of Work; Issue Solicitation

Facility Manager

Determine Qualified Firms; Develop Initial Ranking (Select Short List of A-E Firms)

Conduct Oral Presentations and Interviews with Short-listed Firms

Establish Competitive Range; Select A-E Firms; Issue Requests for Proposal

Evaluate Technical and Cost

Negotiate Hours; Type and Discipline; Direct and Indirect Costs; Period of Performance

Award Design Contract; Issue Notice to Proceed

C

Go to Engineering and Design Process
See Figure A-3

Designer; A-E Firm

Prepare and Submit Qualification Statements

Develop Project-Specific Proposal(s)

Capital Facilities Business Processes

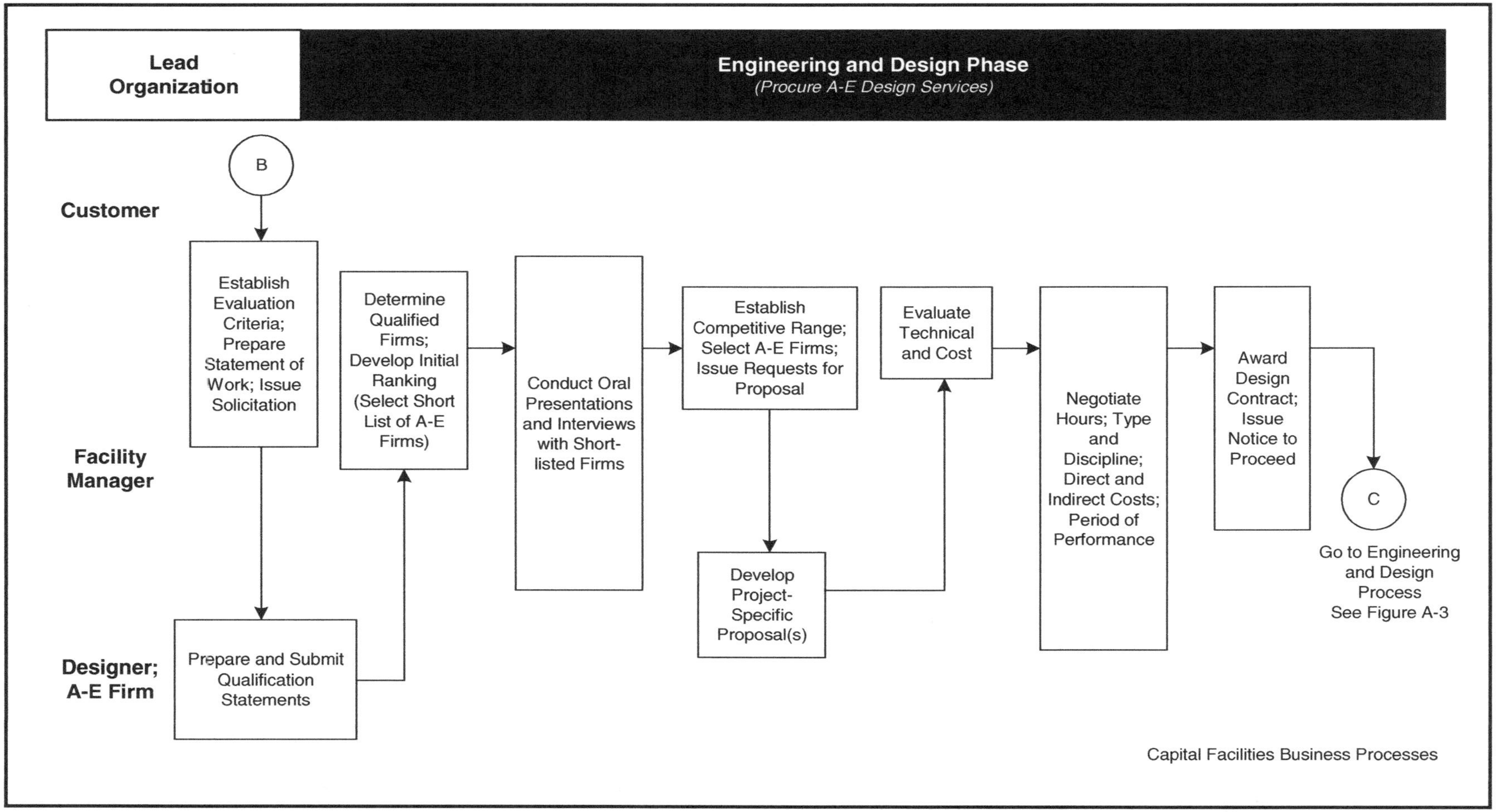

Figure A-5. Typical Business Process of Capital Facilities Project—Procure Construction Services

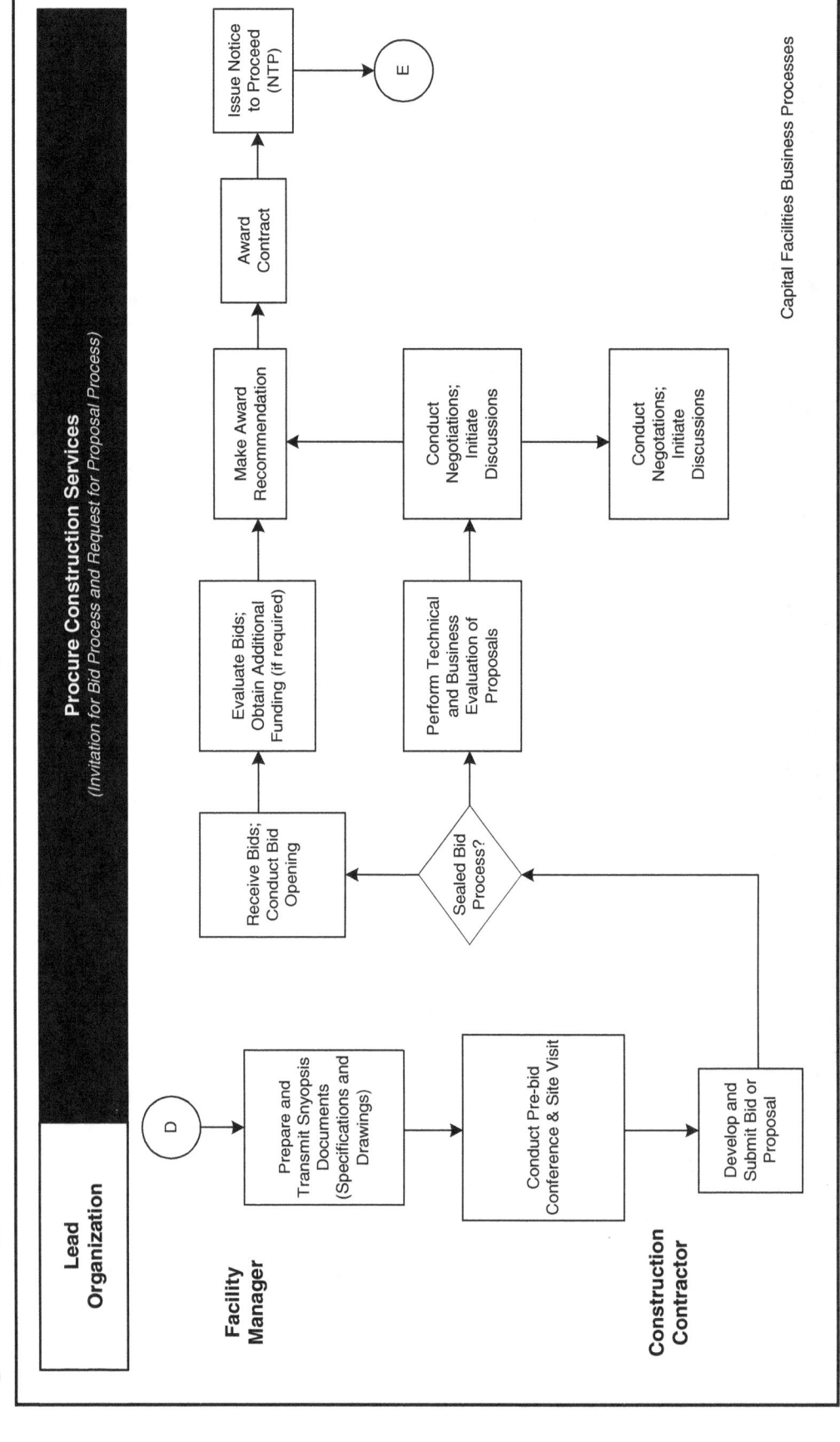

Figure A-6. Typical Business Process of Capital Facilities Project—Construction Phase

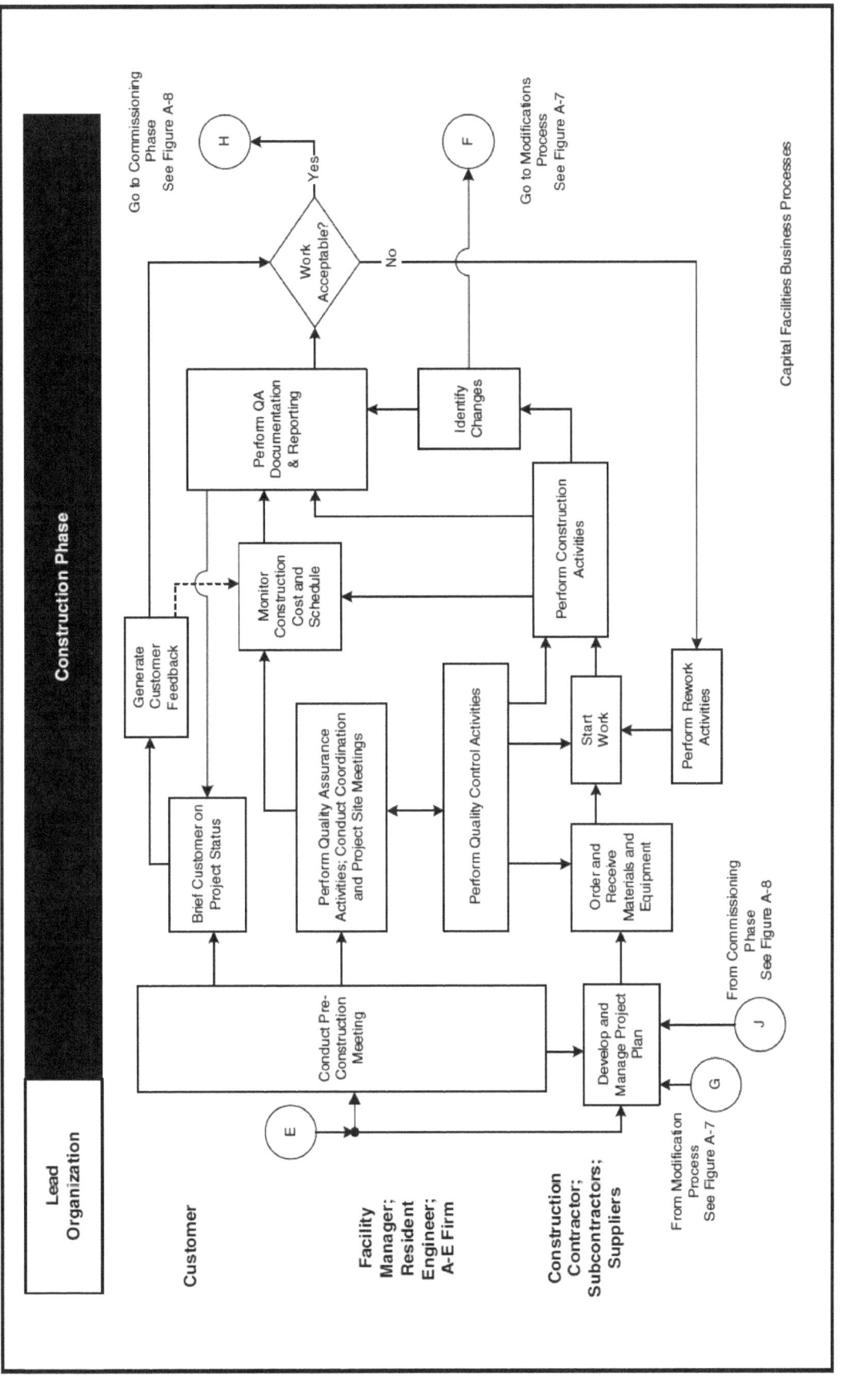

Figure A-7. Typical Business Process of Capital Facilities Project—Construction Phase: Modifications Process

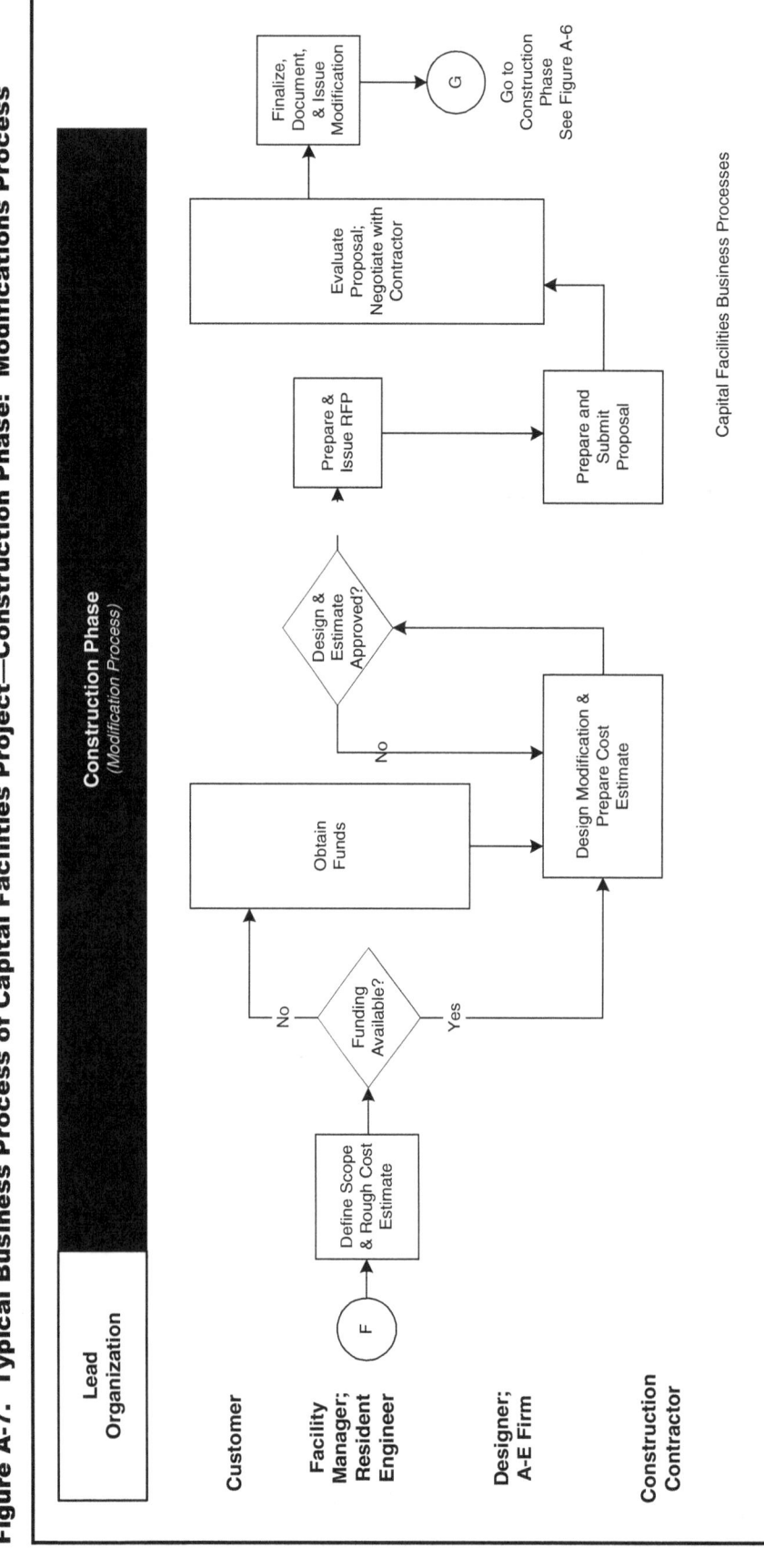

Figure A-8. **Typical Business Process of Capital Facilities Project—Commissioning and Close-out Phase**

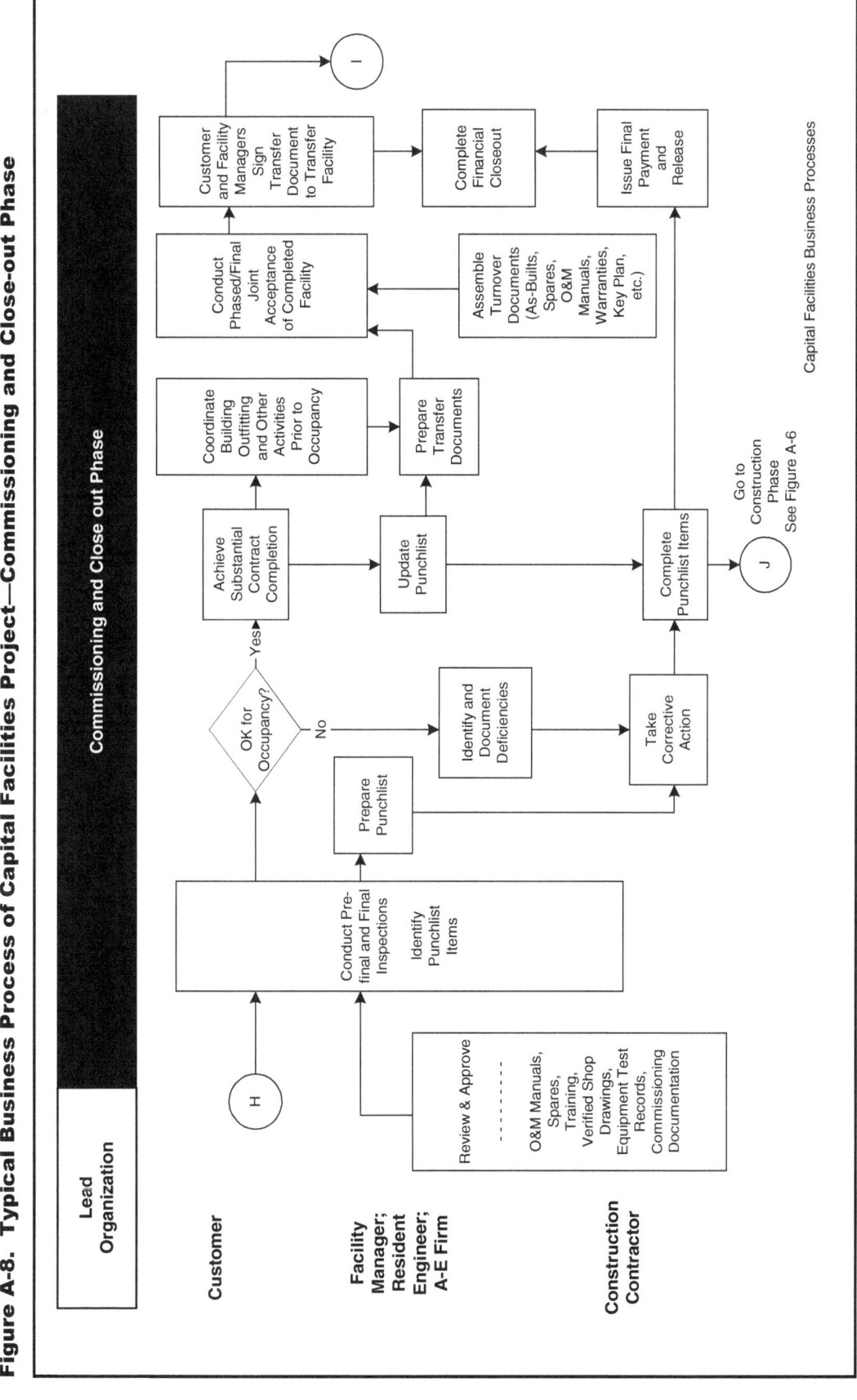

Figure A-9. Typical Business Process of Capital Facilities Project—Operations and Maintenance Phase

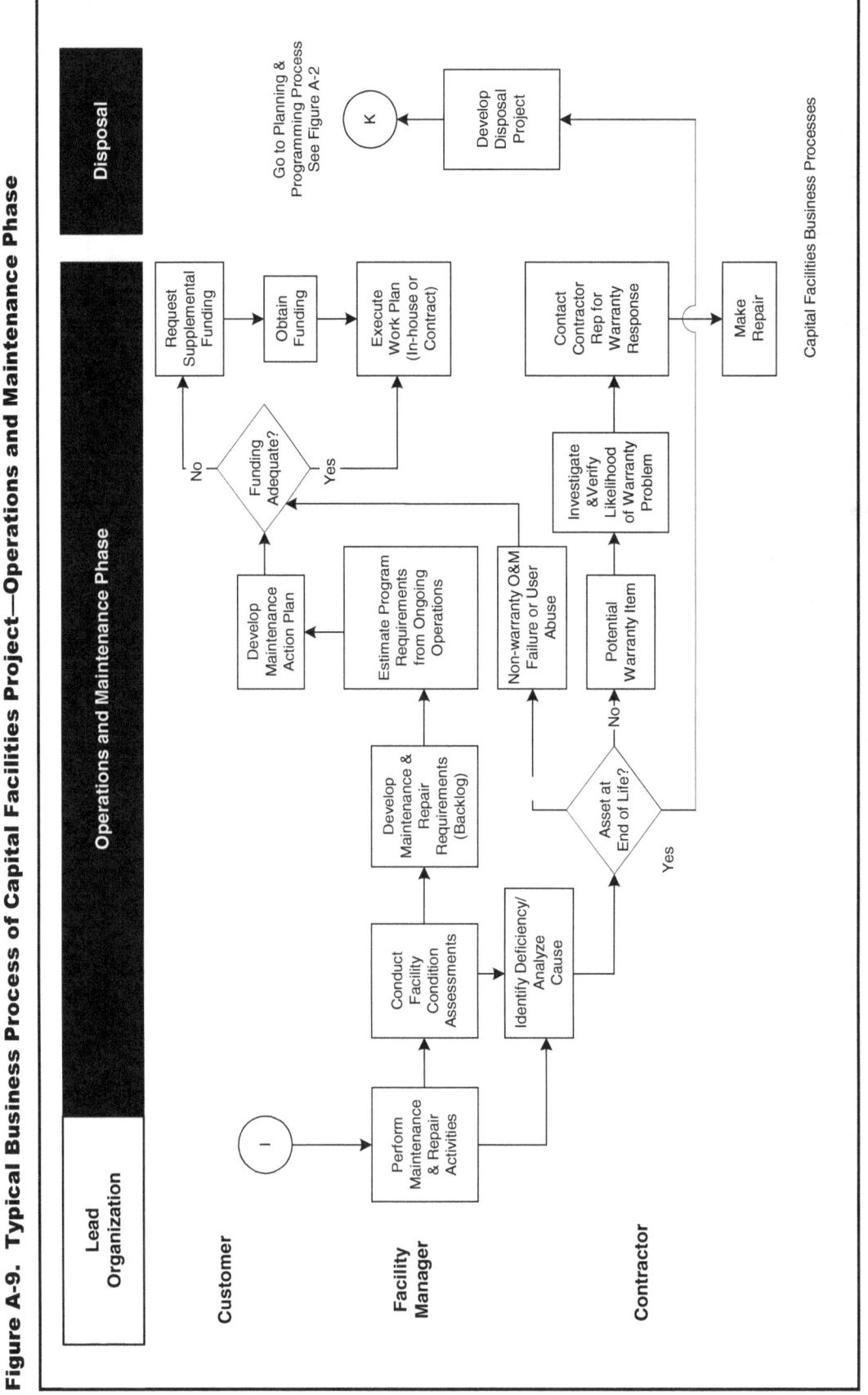

Source: LMI.

Appendix B:
Survey Instruments

Appendix B presents the survey instruments used to collect quantitative data. These surveys were also replicated on the Internet at https://consint.rti.org.

B-1 Owners and Operators Survey Instrument

B-2 General Contractors Survey Instrument

B-3 Specialty Fabricators and Suppliers Survey Instrument

B-4 Architects and Engineers Survey Instrument

B-5 CAD/CAM/CAE/PDM/ERP Software Vendors Survey Instrument

B-1. Owners and Operators Survey Instrument

On behalf of the National Institute of Standards and Technology's (NIST's) Advanced Technology Program (ATP) and Building and Fire Research Laboratory (BFRL), RTI International and Logistics Management Institute (LMI) are conducting a cost analysis of inadequate interoperability in information exchange and management in the capital facilities industry. The goal of the study is to quantify the cost of inefficient information management and data exchange on industry stakeholders, including owners, architects, engineers, constructors, and suppliers involved in the life cycle of commercial, institutional, and industrial facilities.

Examples of these costs include those arising from the software maintenance expenses and labor associated with multiple design systems, the value of manpower required for data translation or reentry, redundant paper and software systems, and investment in third-party interoperability solutions.

Costs may also be generated through design corrections and revisions due to use of incorrect information; the value of manpower expended in the search for, and provision and validation of, redundant paper-based information; and information-access-related project delays.

As a member of the capital facilities supply chain, you have unique insights into the issues associated with inadequate interoperability in the capital facilities life cycle. The information you provide will enable NIST BFRL and industry to identify the impact of inadequate interoperability and plan future research and development efforts in the realm of interoperability.

Please use your experience in the capital facilities industry to answer this brief questionnaire. In addition, please feel free to collaborate with colleagues in your organization to answer the questions. We anticipate that the survey will take approximately 20 to 30 minutes to complete.

The questionnaire you are about to complete is located on a secure server using 128-bit encryption. You will create your own unique user ID and password, which you may share with your colleagues if you decide to make responding to the survey a group effort. In addition, the information you provide is confidential and will only be used in aggregate with responses from other companies in the industry. Your individual response will not be disclosed to any third party, including NIST.

If you have questions, please feel free to contact Alan O'Connor at (919) 541-7186 (oconnor@rti.org) or Mike Gallaher at (919) 541-5935 (mpg@rti.org).

Thank you for your assistance with this important NIST research study.

1. Respondent Identification

Company Name: _____

Mailing Address: _____

Contact Name: _____

Title: _____

Phone Number: _____

E-mail: _____

Is the information in this questionnaire specific to your division, or is it for the entire company or governmental agency?

☐ Division ☐ Company/Agency

[Hereafter your company, division, or agency will be referred to as your "organization."]

Approximately how many employees are in your organization? _____ Employees

What are your organization's capital facilities life-cycle management responsibilities?

_____

_____

_____

2. Capital Facilities Stock Under Construction and Management

These questions ask you to provide some measure of the scale of your organization's average annual capital facilities management activities. This information will allow us to aggregate your responses with those of other organizations.

2.1 In a typical year, in approximately how many *new* capital facilities projects is your organization engaged? _____ Projects

2.2 In a typical year, approximately how many *total* square feet do the above *new* commercial, institutional, and industrial projects represent (excluding petrochemical and utility plants)? _____ Square feet

2.3 What is the distribution of those *new* projects across facility types, by square footage?

Commercial (e.g., office and/or large-scale residential buildings)		Percent
Institutional (e.g., schools and hospitals)		Percent

Industrial (e.g., manufacturing establishments, *except* petrochemical facilities and utilities)		Percent
Total	100%	

2.4   What is your organization's current stock of capital facilities? Please complete the table below using your best estimates of the number and size of your existing facilities and the share of your organizations management activity required by each facility category.

Facility Type	Share of Your Organization's Management Activities (by Labor Hours)	Approximate Number of Facilities	Estimated Total Size of Facilities	Unit
Commercial				Square feet
Institutional				Square feet
Industrial (excluding petrochemical and utility plants)				Square feet
Total	100%			

3.   Design and Construction Life-Cycle Phases

This section explores activities and investments related to information management and exchange during the design, engineering, and construction of capital facilities. These activities also include renovations, modifications, and/or additions to existing facilities.

3.1   CAD/CAM/CAE (CAx) Software Systems

What computer-aided design and engineering systems does your organization use? Please indicate the number of licenses (seats) you have for each system. Please also indicate whether each system is a primary, "in-house" system or a secondary system maintained for coordinating with external parties. A comments field is also provided should you wish to comment on your organization's use of each system.

CAx System Name	Number of Licenses (or Seats)	Is this a preferred in-house (primary) system?	Comments
		☐ Yes ☐ No	
		☐ Yes ☐ No	
		☐ Yes ☐ No	
		☐ Yes ☐ No	
		☐ Yes ☐ No	
		☐ Yes ☐ No	
		☐ Yes ☐ No	
		☐ Yes ☐ No	

The next three questions request the number of employees in your organization who use the CAx software systems listed in the table above. This questionnaire refers to those employees as "CAx users."

3.1.1 How many CAx users are on staff at your organization?   _____ Users

3.1.2 If applicable, what percentage of CAx users use systems that have duplicate capability (i.e., systems that are functional equivalents)?   _____ Percent

3.1.3 Of these users, what is the average amount of time they spend using secondary systems that duplicate the primary system's capability?   _____ Percent

3.2   Interoperability Problems During Design and Construction

This section asks you to reflect on the impact interoperability problems have on your organization's work load during the first two life-cycle phases. The first subsection asks about activities that occur prior to commencing construction. The second subsection asks questions about activities undertaken during the construction phase. Some questions are repeated in both Sections 3.2.1 and 3.2.2; it is important to respond to each question according to activities occurring during the specified time frame only.

3.2.1 Interoperability Problems *Before* Construction Begins

3.2.1.1 Are the responses to this section to be provided on an annual basis for all projects or for an average project?

☐ All projects   ☐ Per project

3.2.1.2 Manual Reentry

    a. Do your employees ever manually reenter information from *paper-based* design and engineering planning information sources into your in-house *electronic* systems?

       ☐ No

       ☐ Yes, requiring about _____ man-hours per month

    b. Do your employees ever manually transfer information from *paper-based* design and engineering planning information sources into your in-house *paper-based* systems?

       ☐ No

       ☐ Yes, requiring about _____ man-hours per month

    c. Do your employees ever manually reenter information from *electronic* design and engineering planning information sources into your in-house *electronic* systems?

       ☐ No

       ☐ Yes, requiring about _____ man-hours per month

3.2.1.3 Do employees require a measurable amount of time to verify that they are reviewing the correct version of either electronic files or paper designs?

    ☐ No

    ☐ Yes, requiring about _____ man-hours per month

3.2.1.4 Managing Requests for Information (RFIs)

    a. How many RFIs does your organization make *before* construction commences on an average project annually?

       _____ RFIs

b. How many man-hours are required to assemble and
execute each RFI, on average?

_____ Man-hours

c. How long does it take, on average, to receive a
satisfactory response to an RFI?

Business
_____ days

3.2.2 Interoperability Problems *During* Construction

3.2.2.1 Manual Reentry

a. Do your employees ever manually reenter as-built information from *paper-based*
design and engineering planning changes into your *electronic* systems?

☐ No

☐ Yes, requiring about _____ man-hours per month

b. Do your employees ever manually transfer as-built information from *paper-based*
design and engineering planning changes into your *paper-based* systems?

☐ No

☐ Yes, requiring about _____ man-hours per month

c. Do your employees ever manually reenter as-built information from as-built
electronic design and engineering planning information sources into your
electronic systems?

☐ No

☐ Yes, requiring about _____ man-hours per month

3.2.2.2 Do employees require a measurable amount of time to verify that they are reviewing the correct version of either electronic files or paper designs?

☐ No

☐ Yes, requiring about _____ man-hours per month

3.2.2.3 Managing RFIs

    a. How many RFIs does your organization make *after*
        construction commences on an average project
        annually?   _____ RFIs

    b. How many man-hours are required to assemble
        and execute each RFI, on average?   _____ Man-hours

    c. How long does it take, on average, to receive a   Business
        satisfactory response to an RFI?   _____ days

4. Operations and Maintenance Phase

This section explores activities and investments related to information management and exchange during the operations and maintenance phase of capital facilities. These questions are specifically related to the facilities management process during this phase.

4.1   In what format is most capital facilities information maintained at your organization? Please also estimate the percentage of facilities management information housed in each format. Note: "Preferred Systems" are the in-house systems that you listed in Question 3.1.

File Format	Percentage of Total Facilities Information	Comment
Paper Files		
Preferred System(s) Electronic Files		
Miscellaneous Electronic Files		
Total	100%	

4.2   Facilities Management Software Systems and Users

4.2.1   Which software systems, if any, does your organization use to manage its capital facilities?

Software System Name	Number of Licenses (or Seats)	Is This a Preferred In-House (primary) System?	Comments
		☐ Yes ☐ No	
		☐ Yes ☐ No	
		☐ Yes ☐ No	
		☐ Yes ☐ No	
		☐ Yes ☐ No	

4.2.2   Facilities Management Software Systems Users

4.2.2.1 How many users on staff at your organization use the systems listed in Question 5.2.1?

_____ Users

4.2.2.2 If applicable, what percentage of those users use systems that have duplicate capability?

_____ Percent

4.2.2.3 Of these users, what is the average amount of time they spend using secondary systems that duplicate the primary system's capability?

_____ Percent

4.3   Facilities Management Staff

The following questions request some information about your facilities management engineers and employees and the amount of time they spend searching for, retrieving, and validating information.

4.3.1   How many operations and maintenance engineers are on staff at your organization?

_____ Employees

4.3.2 What percentage of these operations and maintenance engineers' time is spent searching for and validating design and engineering plan paper archives?

_____ Percent

4.3.3 What percentage of these operations and maintenance engineers' time is spent accessing legacy and miscellaneous electronic files and making them readable by in-house systems?

_____ Percent

4.3.4 What percentage of these operations and maintenance engineers' time is spent waiting for others to provide them with the information needed to perform operations and maintenance tasks?

_____ Percent

4.3.5 If these operations and maintenance engineers had access to that information electronically when they needed it, by what percentage do you estimate their time spent searching for and validating information from paper archives could be reduced?

_____ Percent

4.4 In a typical year, are there incidences when operations and maintenance activities are re-performed because employees were proceeding with the incorrect version of the paper or electronic design and/or engineering files?

☐ No

☐ Yes

　　☐ Occurring about _____ number of times per year

　　☐ Requiring about _____ man-hours per incident

　　☐ Scrapping about _____ dollars' worth of materials per incident

5. Delay Costs Associated with Interoperability Problems and Efforts to Reduce the Occurrence of Those Problems

5.1 In general, what types of delays has your organization experienced because of interoperability problems? What types of costs were associated with those delays?

_____

_____

_____

5.2 These questions ask about your organization's investments in data translation systems to reduce the incidence of poor CAx file transfer. They also ask about your internal research and development activities to reduce interoperability problems, as well as participation in industry consortia aiming to improve interoperability.

5.2.1 If your organization uses data translators licensed from a third-party software vendor, what are the approximate total *annual* licensing fees associated with those translators?

_____ Dollars

5.2.2 If your organization uses third-party data translation and interoperability solutions providers, what is the approximate *annual* cost of those services?

_____ Dollars

5.2.3 Has your firm invested in internal research and development in data translation and interoperability solutions? If yes, approximately how many dollars and/or man-hours are devoted to that activity *annually*?

_____ Dollars

and/or

_____ Man-hours

5.2.4 If your organization participates in industry consortia cooperating on interoperability issues, what is the approximate *annual* cost of participation?

_____ Dollars

and/or

_____ Man-hours

6. Business Process Systems

This section asks about the manpower employed in supporting the business systems that are used through out the capital facilities life cycle supply chain. To simplify responding to this section, the question is presented in table form. For each business process, please indicate whether your organization uses a software system to manage that process. Please also provide the number of full-time equivalent (FTE) employees engaged in that activity. Finally, estimate the approximate reduction in labor effort that could be achieved if information management systems were fully electronic and interoperable internally and externally with clients and teaming partners.

Business Process	Dedicated Software System Used?	Approximate Number of FTEs Engaged in This Activity	Percentage Labor Reduction That Could Be Achieved If Process Were Fully Electronic and/or Interoperable
Cost Estimation	☐ Yes ☐ No		
Document Management	☐ Yes ☐ No		
Enterprise Resource Planning	☐ Yes ☐ No		
Facility Planning and Scheduling	☐ Yes ☐ No		
Facility Simulation	☐ Yes ☐ No		
Information Requests	☐ Yes ☐ No		
Inspection and Certification	☐ Yes ☐ No		
Maintenance Planning and Management	☐ Yes ☐ No		
Materials Management	☐ Yes ☐ No		
Procurement	☐ Yes ☐ No		
Product Data Management	☐ Yes ☐ No		
Project Management	☐ Yes ☐ No		
Start-up and Commissioning	☐ Yes ☐ No		

7.   Comments

Would you like to share other comments about interoperability issues in the capital facilities supply chain? If so, please do so in the space below.

_____

_____

_____

_____

Are you available for further comment about interoperability issues in the capital facilities supply chain?

☐ Yes
☐ No

Please indicate below if you would like to receive a copy of the final report for this analysis. A PDF file will be emailed to you once it has been released by NIST BFRL.

☐ Yes, please email me a copy
☐ No

Thank you!

B-2. General Contractors Survey Instrument

On behalf of the National Institute of Standards and Technology's (NIST's) Advanced Technology Program (ATP) and Building and Fire Research Laboratory (BFRL), RTI International and Logistics Management Institute (LMI) are conducting a cost analysis of inadequate interoperability in information exchange and management in the capital facilities industry. The goal of the study is to quantify the cost of inefficient information management and data exchange on industry stakeholders, including owners, architects, engineers, constructors, and suppliers involved in the life cycle of commercial, institutional, and industrial facilities.

Examples of these costs include those arising from

the software maintenance expenses and labor associated with multiple design systems,

the value of manpower required for data translation or reentry,

redundant paper and software systems, and

investment in third-party interoperability solutions.

Costs may also be generated through

design corrections and revisions due to use of incorrect information;

the value of manpower expended in the search for, and provision and validation of, redundant paper-based information; and

information-access-related project delays.

As a member of the capital facilities supply chain, you have unique insights into the issues associated with inadequate interoperability in the capital facilities life cycle. The information you provide will enable NIST BFRL and industry to identify the impact of inadequate interoperability and plan future research and development efforts in the realm of interoperability.

Please use your experience in the capital facilities industry to answer this brief questionnaire. In addition, please feel free to collaborate with colleagues in your organization to answer the questions. We anticipate that the survey will take approximately 20 to 30 minutes to complete.

The questionnaire you are about to complete is located on a secure server using 128-bit encryption. You will create your own unique user ID and password, which you may share with your colleagues if you decide to make responding to the survey a group effort. In addition, the information you provide is confidential and will only be used in aggregate with responses from other companies in the industry. Your individual response will not be disclosed to any third party, including NIST.

If you have questions, please feel free to contact Alan O'Connor at (919) 541-7186 (oconnor@rti.org) or Mike Gallaher at (919) 541-5935 (mpg@rti.org).

Thank you for your assistance with this important NIST research study.

1.   Respondent Identification

Company Name: _____

Mailing Address: _____

Contact Name: _____

Title: _____

Phone Number: _____

E-mail: _____

Is the information in this questionnaire specific to your division, or is it for the entire company or governmental agency?

☐ Division   ☐ Company/Agency

[Hereafter your company, division, or agency will be referred to as your "organization."]

Approximately how many employees are in your organization?   _____ Employees

What are your organization's capital facilities life-cycle management responsibilities?

_____

_____

_____

2.   Annual Capital Facilities Construction Activities

These questions ask you to provide some measure of the scale of your organization's average annual capital facilities construction activities. This information will allow us to aggregate your response with those of other organizations.

2.1 In a typical year, in approximately how many capital
facilities projects is your organization engaged?

_____ Projects

2.2 In a typical year, approximately how many *total*
square feet do commercial, institutional, and industrial
projects represent (excluding petrochemical and utility
plants)?

_____ Square feet

In a typical year, what is the approximate total
capacity of your petrochemical and utility projects?
Please specify your unit of measure.

_____ (Unit)

2.3 How long does it take to complete construction
activities for a typical project, on average?

_____ Months

2.4 What is the distribution of those projects across facility types, by square footage?

Commercial (e.g., office and/or large-scale residential buildings)		Percent
Institutional (e.g., schools and hospitals)		Percent
Industrial (e.g., manufacturing establishments, *except* petrochemical facilities and utilities)		Percent
Total	100%	

3.   Information Technology Systems and Support

This section explores your organization's investments in and use of software systems to support
your business relationships with clients and teaming partners.

3.1   CAD/CAM/CAE (CAx) Systems

What software systems, such as AutoCAD and MicroStation, does your organization use in its
construction activities for capital facilities projects? Please also indicate if a system duplicates
the capability of your preferred "in-house" system. For example, if AutoCAD is your preferred
system, but your organization also maintains MicroStation, enter MicroStation under "CAx
System Name" but also indicate MicroStation in the third column for your AutoCAD record.

CAx System Name	Number of Licenses (or Seats)	Maintained *Secondary* System with Comparable Capability	Comments

The next three questions request the number of employees in your organization who use the CAx systems listed in Question 3.1. This questionnaire refers to those employees as "CAx users."

3.1.1 How many CAx users are on staff at your organization?   _____ Users

3.1.2 If applicable, what percentage of CAx users use systems that have duplicate capability (i.e., systems that are functional equivalents)?   _____ Percent

3.1.3 Of these users, what is the average amount of time they spend using secondary systems that duplicate the primary system's capability?   _____ Percent

3.2   Data Translation Systems and Interoperability Research

These questions ask about your organization's investments in data translation systems to reduce the incidence of poor CAx file transfer. They also ask about your internal research and development activities to reduce interoperability problems, as well as participation in industry consortia aiming to improve interoperability.

3.2.1 If your organization uses data translators licensed from a third-party software vendor, what are the approximate total *annual* licensing fees associated with those translators?

_____ Dollars

3.2.2 If your organization uses third-party data translation and interoperability solutions providers, what is the approximate *annual* cost of those services?

_____ Dollars

3.2.3 Has your firm invested in internal research and development in data translation and interoperability solutions? If yes, approximately how many man-hours are devoted to that activity *annually*?

_____ Man-hours

3.2.4 If your organization participates in industry consortia cooperating on interoperability issues, what is the approximate *annual* cost of membership and/or donated labor hours for participation?

_____ Dollars

and/or

Man-hours

_____

4. Interoperability Problems

This section asks you to reflect on the impact interoperability problems have on your organization's work load. The first subsection asks about activities that occur prior to commencing construction. The second subsection asks questions about activities undertaken during the construction phase. Some questions are repeated in both Sections 4.1 and 4.2; it is important to respond to each question according to activities occurring during the specified time frame only.

4.1 Interoperability Problems *Before* Construction Commences

4.1.1 Are the responses to this section to be provided on an annual basis for all projects or for an average project?

☐ All projects   ☐ Per project

4.1.1.1 Do your employees ever manually reenter information from *paper-based* design and engineering planning information sources into your in-house *electronic* systems?

☐ No

☐ Yes, requiring about _____ man-hours per month

4.1.1.2 Do your employees ever manually transfer information from *paper-based* design and engineering planning information sources into your in-house *paper-based* systems?

☐ No

☐ Yes, requiring about _____ man-hours per month

4.1.1.3 Do your employees ever manually reenter information from *electronic* design and engineering planning information sources into your in-house *electronic* systems?

☐ No

☐ Yes, requiring about _____ man-hours per month

4.1.2 Do employees require a measurable amount of time to verify that they are working with the correct version of either electronic files or paper designs?

☐ No

☐ Yes, requiring about _____ man-hours per month

4.1.3 Managing Requests for Information (RFIs)

4.1.3.1 How many RFIs does your organization make *before* construction commences on an average project annually?

_____ RFIs

4.1.3.2 How many man-hours are required to assemble and execute each RFI, on average?

_____ Man-hours

4.1.3.3 How long does it take, on average, to receive a satisfactory response to an RFI?

_____ Business days

4.2   Interoperability Problems *During* Construction

4.2.1 Do your employees ever manually reenter as-built information from *paper-based* design and engineering planning changes into *electronic* systems for delivery to teaming partners and owners?

☐ No

☐ Yes, requiring about _____ man-hours per month

4.2.2 Do your employees ever manually transfer as-built information from *paper-based* design and engineering planning changes into *paper-based* systems for delivery to teaming partners and owners?

☐ No

☐ Yes, requiring about _____ man-hours per month

4.2.3 Do your employees ever manually reenter as-built information from as-built *electronic* design and engineering planning information sources into *electronic* systems for delivery to teaming partners and owners?

☐ No

☐ Yes, requiring about _____ man-hours per month

4.2.4 Do employees require a measurable amount of time to verify that they are working with the correct version of either electronic files or paper designs?

☐ No

☐ Yes, requiring about _____ man-hours per month

4.2.5 Managing Requests for Information (RFIs)

4.2.5.1 How many RFIs does your organization make *after* construction commences on an average project annually?

_____ RFIs

4.2.5.2 How many man-hours are required to assemble and execute each RFI, on average?

_____ Man-hours

4.2.5.3 How long does it take, on average, to receive a satisfactory response to an RFI?

_____ Business days

4.2.6 In a typical year, are there incidences when construction set in place has to be reworked because employees were proceeding with the incorrect version of the paper or electronic design and/or engineering files?

    ☐ No

    ☐ Yes

        ☐ Occurring about   _____ number of times per year

        ☐ Requiring about   _____ man-hours per incident

        ☐ Scrapping about   _____ dollars' worth of materials per incident

4.3 Interoperability Problems *After* Construction Ends

4.3.1 Do employees perform redundant tasks in transferring information to owners and operators after construction is completed, due to software systems that lack interoperability?

    ☐ No

    ☐ Yes, requiring about   _____ man-hours per month

5. Impact of Delays Due to Interoperability Problems

5.1 When construction-related activities are halted because of the submission of RFIs or other information-access related issues, are employees idle during this time?

    ☐ No

    ☐ Yes, idling about   _____ man-hours per month

5.2 If general contractors had access to all the information they needed when they needed it, would the average length of time required to complete a project be reduced?

    ☐ No

    ☐ Yes, about   months   percent

    _____ or   _____

5.3   In general, what other types of delays has your organization experienced because of interoperability problems? What types of costs were associated with those delays?

_____

_____

_____

_____

6.   Business Process Systems

This subsection asks whether your organization uses software systems to support certain business processes in the capital facilities supply chain. To simplify responding to this section, the question is presented in table form. For each business process, please indicate whether your organization uses a software system to facilitate information management. Please also provide the number of full-time equivalent (FTE) employees engaged in that process. Finally, estimate the approximate reduction in labor effort that could be achieved if information management systems were fully electronic and interoperable internally and with clients and teaming partners.

Business Process	Dedicated Software System Used?	Approximate Number of FTEs Engaged in This Activity	Percentage Labor Reduction That Could Be Achieved If Process Were Fully Electronic and/or Interoperable
Accounting	☐ Yes ☐ No		
Cost Estimation	☐ Yes ☐ No		
Document Management	☐ Yes ☐ No		
Enterprise Resource Planning	☐ Yes ☐ No		
Facility Planning and Scheduling	☐ Yes ☐ No		
Facility Simulation	☐ Yes ☐ No		
Information Requests	☐ Yes ☐ No		
Inspection and Certification	☐ Yes ☐ No		
Maintenance Planning and Management	☐ Yes ☐ No		
Materials Management	☐ Yes ☐ No		
Procurement	☐ Yes ☐ No		
Product Data Management	☐ Yes ☐ No		
Project Management	☐ Yes ☐ No		
Start-up and Commissioning	☐ Yes ☐ No		

7.   Comments

Would you like to share other comments about interoperability issues in the capital facilities supply chain? If so, please do so in the space below.

_____

_____

_____

_____

Are you available for further comment about interoperability issues in the capital facilities supply chain?

    ☐ Yes
    ☐ No

Please indicate below if you would like to receive a copy of the final report for this analysis. A PDF file will be emailed to you once it has been released by NIST BFRL.

    ☐ Yes, please email me a copy
    ☐ No

Thank you!

B-3. Specialty Fabricators and Suppliers Survey Instrument

On behalf of the National Institute of Standards and Technology's (NIST's) Advanced Technology Program (ATP) and Building and Fire Research Laboratory (BFRL), RTI International and Logistics Management Institute (LMI) are conducting a cost analysis of inadequate interoperability in information exchange and management in the capital facilities industry. The goal of the study is to quantify the cost of inefficient information management and data exchange on industry stakeholders, including owners, architects, engineers, constructors, and suppliers involved in the life cycle of commercial, institutional, and industrial facilities.

Examples of these costs include those arising from

the software maintenance expenses and labor associated with multiple design systems,

the value of manpower required for data translation or reentry,

redundant paper and software systems, and

investment in third-party interoperability solutions.

Costs may also be generated through

design corrections and revisions due to use of incorrect information;

the value of manpower expended in the search for, and provision and validation of, redundant paper-based information; and

information-access-related project delays.

As a member of the capital facilities supply chain, you have unique insights into the issues associated with inadequate interoperability in the capital facilities life cycle. The information you provide will enable NIST BFRL and industry to identify the impact of inadequate interoperability and plan future research and development efforts in the realm of interoperability.

Please use your experience in the capital facilities industry to answer this brief questionnaire. In addition, please feel free to collaborate with colleagues in your organization to answer the questions. We anticipate that the survey will take approximately 20 to 30 minutes to complete.

The questionnaire you are about to complete is located on a secure server using 128-bit encryption. You will create your own unique user ID and password, which you may share with your colleagues if you decide to make responding to the survey a group effort. In addition, the information you provide is confidential and will only be used in aggregate with responses from other companies in the industry. Your individual response will not be disclosed to any third party, including NIST.

If you have questions, please feel free to contact Alan O'Connor at (919) 541-7186 (oconnor@rti.org) or Mike Gallaher at (919) 541-5935 (mpg@rti.org).

Thank you for your assistance with this important NIST research study.

1.   Respondent Identification

Company Name: _____

Mailing Address: _____

Contact Name: _____

Title: _____

Phone Number: _____

E-mail: _____

Is the information in this questionnaire specific to your division, or is it for the entire company or governmental agency?

☐   Division   ☐   Company/Agency

[Hereafter your company, division, or agency will be referred to as your "organization."]

Approximately how many employees are in your organization?   _____ Employees

What are your organization's capital facilities life-cycle management responsibilities?

_____

_____

_____

2. Annual Capital Facilities Specialty and Fabrication and Supply Activities

These questions ask you to provide some measure of the scale of your organization's average annual capital facilities fabrication and supply activities. This information will allow us to aggregate your response with those of other organizations.

2.1 In a typical year, in approximately how many capital facilities projects is your organization engaged?

_____ Projects

2.2 In a typical year, approximately how many *total* square feet do commercial, institutional, and industrial projects represent (excluding petrochemical and utility plants)?

_____ Square feet

In a typical year, what is the approximate total capacity of your petrochemical and utility projects? Please specify your unit of measure.

_____ (Unit)

2.3 What is the distribution of those projects across facility types, by square footage?

Commercial (e.g., office and/or large-scale residential buildings)		Percent
Institutional (e.g., schools and hospitals)		Percent
Industrial (e.g., manufacturing establishments, *except* petrochemical facilities and utilities)		Percent
Total	100%	

3. Information Technology Systems and Support

This section explores your organization's investments in and use of software systems, if any, to support your internal operations as well as your relationships with clients and teaming partners.

3.1 CAD/CAM/CAE (CAx) Systems

What software systems, such as AutoCAD and MicroStation, does your organization use in its specialty construction and fabrication activities for capital facilities projects? Please also indicate if a system duplicates the capability of your preferred "in-house" system. For example, if AutoCAD is your preferred system, but your organization also maintains MicroStation, enter MicroStation under "CAx System name" but also indicate MicroStation in the third column for your AutoCAD record.

CAx System Name	Number of Licenses (or Seats)	Maintained *Secondary* System with Comparable Capability	Comments

The next three questions request the number of employees in your organization who use the CAx systems listed in Question 3.1. This questionnaire refers to those employees as "CAx users."

3.1.1 How many employees use the systems indicated in the above table?

_____ Users

3.1.2 If applicable, what percentage of CAx users use systems that have duplicate capability (i.e., systems that are functional equivalents)?

_____ Percent

3.1.3 Of these users, what is the average amount of time they spend using secondary systems that duplicate the primary system's capability?

_____ Percent

3.2 Data Translation Systems and Interoperability Research

These questions ask about your organization's investments in data translation systems to reduce the incidence of poor CAx file transfer. They also ask about your internal research and development activities to reduce interoperability problems, as well as participation in industry consortia aiming to improve interoperability.

3.2.1 If your organization uses data translators licensed from a third-party software vendor, what are the approximate total *annual* licensing fees associated with those translators?

_____ Dollars

3.2.2 If your organization uses third-party data translation and interoperability solutions providers, what is the approximate *annual* cost of those services?

_____ Dollars

3.2.3 Has your firm invested in internal research and development in data translation and interoperability solutions? If yes, approximately how many man-hours are devoted to that activity *annually*?

_____ Man-hours

3.2.4 If your organization participates in industry consortia cooperating on interoperability issues, what is the approximate *annual* cost of membership and/or donated labor hours for participation?

_____ Dollars
and/or
_____ Man-hours

4.   Interoperability Problems

The questions in this section ask you to reflect on the impact interoperability problems have on your organization's work load. The first subsection asks about activities that occur prior to commencing construction. The second subsection asks questions about activities undertaken during the construction phase. Some questions are repeated in both Sections 4.1 and 4.2; it is important to respond to each question according to activities occurring during the specified time frame only.

4.1   Interoperability Problems *Before* Construction

4.1.1 Are the responses to this section to be provided on an annual basis for all projects or for an average project?

☐ Total   ☐ Per project

4.1.2 Manual Reentry

4.1.2.1 Do your employees ever manually reenter information from *paper-based* design and engineering planning information sources into your in-house *electronic* systems?

☐ No

☐ Yes, requiring about _____ man-hours per month

4.1.2.2 Do your employees ever manually transfer information from *paper-based* design and engineering planning information sources into your in-house *paper-based* systems?

☐ No

☐ Yes, requiring about _____ man-hours per month

4.1.2.3 Do your employees ever manually reenter information from *electronic* design and engineering planning information sources into your in-house *electronic* systems?

☐ No

☐ Yes, requiring about _____ man-hours per month

4.1.3 Do employees require a measurable amount of time to verify that they are working with the correct version of either electronic files or paper designs?

☐ No

☐ Yes, requiring about _____ man-hours per month

4.1.4 Managing Requests for Information (RFIs)

4.1.4.1 How many RFIs does your organization make *before* construction commences on an average project annually?

_____ RFIs

4.1.4.2 How many man-hours are required to assemble and execute each RFI, on average?

_____ Man-hours

4.1.4.3 How long does it take, on average, to receive a satisfactory response to an RFI?

_____ Business days

4.2 Interoperability Problems *After* Construction Commences

4.2.1 Do your employees ever manually reenter as-built information from *paper-based* design and engineering planning changes into *electronic* systems for delivery to teaming partners and owners?

☐ No

☐ Yes, requiring about _____ man-hours per month

4.2.2 Do your employees ever manually transfer as-built information from *paper* design and engineering planning changes into *paper* systems for delivery to teaming partners and owners?

☐ No

☐ Yes, requiring about   _____   man-hours per month

4.2.3 Do your employees ever manually reenter as-built information from as-built *electronic* design and engineering planning information sources into *electronic* systems for delivery to teaming partners and owners?

☐ No

☐ Yes, requiring about   _____   man-hours per month

4.2.4 Do employees require a measurable amount of time to verify that they are working with the correct version of either electronic files or paper designs?

☐ No

☐ Yes, requiring about   _____   man-hours per month

4.2.5 Managing RFIs

4.2.5.1 How many RFIs does your organization make *after* construction commences on an average project annually?

_____ RFIs

4.2.5.2 How many man-hours are required to assemble and execute each RFI, on average?

_____ Man-hours

4.2.5.3 How long does it take, on average, to receive a satisfactory response to an RFI?

_____ Business days

4.2.6 In a typical year, are there incidences when construction set in place has to be reworked because employees were proceeding with the incorrect version of the paper or electronic design and/or engineering files?

☐ No

☐ Yes

    ☐ Occurring about   _____ number of times per year

    ☐ Requiring about   _____ man-hours per incident

    ☐ Scrapping about   _____ dollars' worth of materials per incident

5. Impact of Delays Due to Interoperability Problems

5.1 When construction-related activities are halted because of the submission of RFIs or other information-access related issues, are employees idle during this time?

☐ No

☐ Yes, idling about _____ man-hours per month

5.2 If employees in your organization had access to all the information they needed when they needed it, would the average length of time required to construct a new facility be reduced?

☐ No

☐ Yes, about _____ months or _____ percent

5.3 In general, what other types of delays has your organization experienced because of interoperability problems? What types of costs were associated with those delays?

_____

_____

_____

_____

6. Business Process Systems

This section asks whether your organization uses software systems to support certain business processes in the capital facilities supply chain. To simplify responding to this section, the question is presented in table form. For each business process, please indicate whether your organization uses a software system to facilitate information management. Please also provide the number of full-time equivalent (FTE) employees engaged in that process. Finally, estimate the approximate reduction in labor effort that could be achieved if information management systems were fully electronic and interoperable internally and with clients and teaming partners.

Business Process	Dedicated Software System Used?	Approximate Number of FTEs Engaged in This Activity	Percentage Labor Reduction That Could Be Achieved If Process Were Fully Electronic and/or Interoperable
Accounting	☐ Yes ☐ No		
Cost Estimation	☐ Yes ☐ No		
Document Management	☐ Yes ☐ No		
Enterprise Resource Planning	☐ Yes ☐ No		
Facility Planning and Scheduling	☐ Yes ☐ No		
Facility Simulation	☐ Yes ☐ No		
Information Requests	☐ Yes ☐ No		
Inspection and Certification	☐ Yes ☐ No		
Maintenance Planning and Management	☐ Yes ☐ No		
Materials Management	☐ Yes ☐ No		
Procurement	☐ Yes ☐ No		
Product Data Management	☐ Yes ☐ No		
Project Management	☐ Yes ☐ No		
Start-up and Commissioning	☐ Yes ☐ No		

7.   Comments

Would you like to share other comments about interoperability issues in the capital facilities supply chain? If so, please do so in the space below.

_____

_____

_____

_____

Are you available for further comment about interoperability issues in the capital facilities supply chain?

☐ Yes
☐ No

Please indicate below if you would like to receive a copy of the final report for this analysis. A PDF file will be emailed to you once it has been released by NIST BFRL.

☐ Yes, please email me a copy
☐ No

Thank you!

B-4. Architects and Engineers Survey Instrument

On behalf of the National Institute of Standards and Technology's (NIST's) Advanced Technology Program (ATP) and Building and Fire Research Laboratory (BFRL), RTI International and Logistics Management Institute (LMI) are conducting a cost analysis of inadequate interoperability in information exchange and management in the capital facilities industry. The goal of the study is to quantify the cost of inefficient information management and data exchange on industry stakeholders, including owners, architects, engineers, constructors, and suppliers involved in the life cycle of commercial, institutional, and industrial facilities.

Examples of these costs include those arising from

the software maintenance expenses and labor associated with multiple design systems,

the value of manpower required for data translation or reentry,

redundant paper and software systems, and

investment in third-party interoperability solutions.

Costs may also be generated through

design corrections and revisions due to use of incorrect information;

the value of manpower expended in the search for, and provision and validation of, redundant paper-based information; and

information-access-related project delays.

As a member of the capital facilities supply chain, you have unique insights into the issues associated with inadequate interoperability in the capital facilities life cycle. The information you provide will enable NIST BFRL and industry to identify the impact of inadequate interoperability and plan future research and development efforts in the realm of interoperability.

Please use your experience in the capital facilities industry to answer this brief questionnaire. In addition, please feel free to collaborate with colleagues in your organization to answer the questions. We anticipate that the survey will take approximately 20 to 30 minutes to complete.

The questionnaire you are about to complete is located on a secure server using 128-bit encryption. You will create your own unique user ID and password, which you may share with your colleagues if you decide to make responding to the survey a group effort.   In addition, the information you provide is confidential and will only be used in aggregate with responses from other companies in the industry. Your individual response will not be disclosed to any third party, including NIST.

If you have questions, please feel free to contact Alan O'Connor at (919) 541-7186 (oconnor@rti.org) or Mike Gallaher at (919) 541-5935 (mpg@rti.org).

Thank you for your assistance with this important NIST research study.

1. Respondent Identification

Company Name: _____

Mailing Address: _____

Contact Name: _____

Title: _____

Phone Number: _____

E-mail: _____

Is the information in this questionnaire specific to your division, or is it for the entire company or governmental agency?

☐ Division   ☐ Company/Agency

[Hereafter your company, division, or agency will be referred to as your "organization."]

Approximately how many employees are in your organization?   _____ Employees

What are your organization's capital facilities life-cycle management responsibilities?

_____

_____

_____

2. Annual Design, Engineering, and Planning Activities

These questions ask you to provide some measure of the scale of your organization's average annual design, engineering, and/or planning activities for capital facilities. This information will allow us to aggregate your response with those of other organizations.

2.1 In a typical year, in approximately how many capital
facilities projects is your organization engaged?

_____ Projects

2.2 In a typical year, approximately how many *total* square
feet do commercial, institutional, and industrial projects
represent (excluding petrochemical and utility plants)?

_____ Square feet

In a typical year, what is the approximate total capacity of
year petrochemical and utility projects? Please specify
your unit of measure.

_____ (Unit)

2.3 What is the distribution of those projects across facility types, by square footage?

Commercial (e.g., office and/or large-scale residential buildings)		Percent
Institutional (e.g., schools and hospitals)		Percent
Industrial (e.g., manufacturing establishments, *except* petrochemical facilities and utilities)		Percent
Total	100%	

3. Information Technology Systems and Support

This section explores your organization's investments in and use of software systems, if any, to
support your internal operations as well as your relationships with clients and teaming partners.

3.1 CAD/CAM/CAE (CAx) Systems

What CAx software systems, such as AutoCAD and MicroStation, does your organization use in
its design and/or engineering work for capital facilities projects? Please also indicate if a system
duplicates the capability of your preferred "in-house" system. For example, if AutoCAD is your
preferred design system, but your organization also maintains MicroStation, enter MicroStation
under "CAx System Name" but also indicate MicroStation in the third column for your AutoCAD
record.

CAx System Name	Number of Licenses (or Seats)	Maintained *Secondary* System with Comparable Capability	Comments

The next three questions request the number of architects, designers, and engineers in your organization who use the CAx systems listed in Question 3.1. This questionnaire refers to those employees as "CAx users."

3.1.1 How many employees use the systems indicated in the above table?

_____ Users

3.1.2 If applicable, what percentage of CAx users use systems that have duplicate capability (i.e., systems that are functional equivalents)?

_____ Percent

3.1.3 Of these users, what is the average amount of time they spend using secondary systems that duplicate the primary system's capability?

_____ Percent

3.2 Data Translation Systems and Interoperability Research

These questions ask about your organization's investments in data translation systems to reduce the incidence of poor CAx file transfer. They also ask about your internal research and development activities to reduce interoperability problems, as well as participation in industry consortia aiming to improve interoperability.

3.2.1 If your organization uses data translators licensed from a third-party software vendor, what are the approximate total *annual* licensing fees associated with those translators?

_____ Dollars

3.2.2 If your organization uses third-party data translation and interoperability solutions providers, what is the approximate *annual* cost of those services?

_____ Dollars

3.2.3 Has your firm invested in internal research and development in data translation and interoperability solutions? If yes, approximately how many man-hours are devoted to that activity *annually*?

Man-
_____ hours

3.2.4 If your organization participates in industry consortia cooperating on interoperability issues, what is the approximate *annual* cost of membership and/or donated labor hours for participation?

_____ Dollars
and/or
Man-
_____ hours

4.   Interoperability Problems

This section asks you to reflect on the impact interoperability problems have on your organization's work load. The first subsection asks about activities during the design, engineering, and planning phase of a new facility. The second subsection asks questions about activities undertaken during the construction phase (i.e., after the final design and engineering plans have been submitted, approved, and implemented). These questions are tailored specifically to your organization's employees tasked with performing facility design, engineering, and or planning work.

4.1   Interoperability Problems *Before* Construction Commences

4.1.1   Are the responses to this section to be provided on an annual basis for all projects or for an average project?

☐ All projects   ☐ Per project

4.1.1.2 Do employees ever manually reenter information from paper-design changes *and/or* electronic design files into your electronic systems? This may occur as a result of alterations based on comments and design changes submitted by owner/operators and teaming partners or poor electronic file transfer.

☐ No
☐ Yes, requiring about _____ man-hours per month

4.1.1.3 Do employees require a measurable amount of time to verify that they are working with the correct version of either electronic files or paper designs?

   ☐ No

   ☐ Yes, requiring about _____ man-hours per month

4.1.2 In a typical year, are there incidences when designs and/or engineering plans have had to be reworked because employees were proceeding with the incorrect version of the paper or electronic files?

   ☐ No

   ☐ Yes

       ☐ Occurring about _____ number of times per year

       ☐ Requiring about _____ hours of rework per incidence

4.2   Interoperability Problems *After* Construction Commences

4.2.1 Do employees ever manually reenter design changes from paper and/or electronic files from general contractors or owner/operators into your electronic systems after construction has commenced? This may occur because of alterations due to construction activity or submission of comments and design changes or poor electronic file transfer.

   ☐ No

   ☐ Yes, requiring about _____ man-hours per month

4.2.2 Do employees require a measurable amount of time to verify that they are working with the correct version of either electronic files or paper designs?

   ☐ No

   ☐ Yes, requiring about _____ man-hours per month

4.2.3 When managing requests for information (RFIs) from either general contractors or owner/operators, do employees spend a measurable amount of time transferring information into either a paper-based format or a new electronic file for delivery to requesting parties?

   ☐ No

   ☐ Yes, requiring about _____ man-hours per month

4.3 Interoperability Problems *After* Construction Ends

4.3.1 Do employees perform redundant tasks in transferring information to owners and operators after construction is completed, due to software systems that lack interoperability?

☐ No

☐ Yes, requiring about _____ man-hours per month

5. Impact of Delays Due to Interoperability Problems

In general, what types of delays has your organization experienced because of interoperability problems? What types of costs were associated with those delays?

_____

_____

6. Business Process Systems

This section asks whether your organization uses software systems to support certain business processes. To simplify responding to this section, the question is presented in table form. For each business process, please indicate whether your organization uses a software system to facilitate information management. Please also provide the number of full-time equivalent (FTE) employees engaged in that process. Finally, estimate the approximate reduction in labor effort that could be achieved if information management systems were fully electronic and interoperable internally and with clients and teaming partners.

Business Process	Dedicated Software System Used?	Approximate Number of FTEs Engaged in This Activity	Percentage Labor Reduction That Could Be Achieved If Process Were Fully Electronic and/or Interoperable
Accounting	☐ Yes ☐ No		
Cost Estimation	☐ Yes ☐ No		
Document Management	☐ Yes ☐ No		
Enterprise Resource Planning	☐ Yes ☐ No		
Facility Planning and Scheduling	☐ Yes ☐ No		
Facility Simulation	☐ Yes ☐ No		
Information Requests	☐ Yes ☐ No		
Materials Management	☐ Yes ☐ No		
Procurement	☐ Yes ☐ No		
Project Management	☐ Yes ☐ No		

7.   Comments

Would you like to share other comments about interoperability issues in the capital facilities supply chain? If so, please do so in the space below.

_____

_____

_____

_____

Are you available for further comment about interoperability issues in the capital facilities supply chain?

☐ Yes
☐ No

Please indicate below if you would like to receive a copy of the final report for this analysis. A PDF file will be emailed to you once it has been released by NIST BFRL.

☐ Yes, please email me a copy
☐ No

Thank you!

B-5. CAD/CAM/CAE/PDM/ERP Software Vendors Survey Instrument

On behalf of the National Institute of Standards and Technology's (NIST's) Advanced Technology Program (ATP) and Building and Fire Research Laboratory (BFRL), RTI International and Logistics Management Institute (LMI) are conducting a cost analysis of inadequate interoperability in information exchange and management in the capital facilities industry. The goal of the study is to quantify the cost of inefficient information management and data exchange on industry stakeholders, including owners, architects, engineers, constructors, and suppliers involved in the life cycle of commercial, institutional, and industrial facilities.

NIST BFRL's aim is to measure the cost associated with the inadequate interoperability in both information exchange and management. Examples of these costs include those arising from the purchase and labor associated with the value of labor lost due to data translation or reentry, redundant paper systems, and investment in third-party interoperability solutions. Such costs may also be generated through design changes due to initial use of incorrect information; value of labor lost in the search for, provision, and validation of redundant paper-based information; and information-access-related project delays.

As a member of the industry that produces the software used in the design, engineering, and facilities management operations of the industry we are investigating, you have unique insights into the state of intersystem connectivity of CAD, CAM, CAE, PDM, and/or ERP software. The information you provide will help NIST better assess the costs of inadequate interoperability and the research and development needs, thereby allowing NIST to channel future investments toward projects that best meet those needs.

Please answer the questions in the attached questionnaire with reference to your CAD, CAM, CAE, PDM, or ERP software products. In addition, please feel free to collaborate with colleagues in answering the questions. *The information you provide is confidential and will only be used in aggregate with responses from other companies in the industry. Your individual response will not be disclosed to any third party, including NIST.*

If you have questions, please feel free to contact Alan O'Connor at (919) 541-7186 (oconnor@rti.org) or Mike Gallaher at (919) 541-5935 (mpg@rti.org).

Thank you for assistance with this important NIST research study.

1.   Company Identification

Company Name: _____

Mailing Address: _____

_____

Contact Name: _____

Title: _____

Phone Number: _____

E-mail: _____

2.   CAD/CAM/CAE/PDM/ERP Product information

2.1   Please list your company's CAD, CAM, CAE, PDM, or ERP software packages and specialty products below that are used by the capital facilities industry (i.e., for the design and engineering, and facilities management of commercial, institutional, and industrial buildings).

_____

_____

_____

_____

_____

_____

2.2   Do the CAD, CAM, CAE, PDM, or ERP software programs your firm markets currently offer any neutral format or intersystem functionality, or will they in the near future?

☐ Yes. In which year(s) did or will these programs first include neutral file format capability and/or intersystem connectivity?

_____

_____

☐ No. *(End survey)*

_____

2.3   Which neutral file formats do your software systems currently support?

_____

_____

2.4   With which systems are your software systems interoperable? Are these systems predominantly within your firm's product family or do they also have connectivity with other firms' offerings?

_____

_____

3.   The Cost of Developing Neutral File Format Functionality or Intersystem Connectivity

3.1   Was your company involved in the administrative process to develop the standards for neutral file format functionality or intersystem connectivity, in developing new technologies and tools, or in supporting demonstrations or certification testing?

☐Yes. Over what time frame did you participate and what were your approximate annual expenditures in terms of person-months?

Activities	Time Period Involved (Example: 1995 to 2001)	Average Annual Expenditures (person-months/year)
Standards development process *(Example: Attended meeting or reviewed draft standards)*		
Software development tools and testing tools *(Example: Supported the development of languages or libraries)*		
Demonstration and certification services *(Example: Participated in the AutoSTEP project or other implementation forums)*		

☐ No. Our company was not involved in these activities.

3.2 What were your company's total expenditures to integrate neutral file formats and/or develop intersystem connectivity into your CAD, CAM, CAE, PDM, or ERP systems? ***(Choose one)***

_____ Dollars

or

_____ Labor (person-months)

4. Comments

4.1 Please provide any additional comments that would help us evaluate the cost of integrating neutral file format and/or intersystem connectivity into your CAD, CAM, CAE, PDM, or ERP software products.

_____

_____

_____

_____

_____

_____

Thank you for your participation.

Please indicate below if you would like to receive a copy of the final report.

☐ Yes, please send a copy
☐ No

Are you available for further comment about interoperability issues in the capital facilities supply chain?   _____

Appendix C:
Wage Rates

Table C-1. Wage Rates for Architects and Engineers

Group	Labor Category	BLS Title	Source	Wage (2002)	Loaded (2002)
Arch	CAx User	Architects, Except Landscape and Naval	http://www.bls.gov/oes/2002/naics4_541300.htm	29.88	59.76
Arch	Design Support Specialist	Architects, Except Landscape and Naval	http://www.bls.gov/oes/2002/naics4_541300.htm	29.88	59.76
Arch	Software Support Specialist	Computer Support Specialists	http://www.bls.gov/oes/2002/naics4_541300.htm	20.84	41.68
Arch	Network and Systems Administrators	Network and Computer Systems Administrators	http://www.bls.gov/oes/2002/naics4_541300.htm	27.91	55.82
Arch	Cost Estimation	Cost Estimators	http://www.bls.gov/oes/2002/naics4_541300.htm	29.43	58.86
Arch	Document Management	Executive Secretaries and Administrative Assistants	http://www.bls.gov/oes/2002/naics4_541300.htm	17.67	35.34
Arch	Enterprise Resource Planning	Management Analysts	http://www.bls.gov/oes/2002/naics4_541300.htm	35.11	70.22
Arch	Facility Planning and Scheduling	Civil Engineers	http://www.bls.gov/oes/2002/naics4_541300.htm	30.53	61.06
Arch	Facility Simulation	Civil Engineers	http://www.bls.gov/oes/2002/naics4_541300.htm	30.53	61.06
Arch	Information Requests	Executive Secretaries and Administrative Assistants	http://www.bls.gov/oes/2002/naics4_541300.htm	17.67	35.34
Arch	Materials Management	Production, Planning, and Expediting Clerks	http://www.bls.gov/oes/2002/naics4_541300.htm	18.73	37.46
Arch	Procurement	Purchasing Agents, Except Wholesale, Retail, and Farm Products	http://www.bls.gov/oes/2002/naics4_541300.htm	25.78	51.56
Arch	Project Management	Architects, Except Landscape and Naval	http://www.bls.gov/oes/2002/naics4_541300.htm	29.88	59.76
Arch	Accountant	Accountants and Auditors	http://www.bls.gov/oes/2002/naics4_541300.htm	25.49	50.98
Eng	CAx User	Civil Engineers	http://www.bls.gov/oes/2002/naics4_541300.htm	30.53	61.06
Eng	Design Support Specialist	Civil Engineers	http://www.bls.gov/oes/2002/naics4_541300.htm	30.53	61.06

(continued)

Table C-1. Wage Rates for Architects and Engineers (continued)

Group	Labor Category	BLS Title	Source	Wage (2002)	Loaded (2002)
Eng	Software Support Specialist	Computer Support Specialists	http://www.bls.gov/oes/2002/naics4_541300.htm	20.84	41.68
Eng	Network and Systems Administrators	Network and Computer Systems Administrators	http://www.bls.gov/oes/2002/naics4_541300.htm	27.91	55.82
Eng	Cost Estimation	Cost Estimators	http://www.bls.gov/oes/2002/naics4_541300.htm	29.43	58.86
Eng	Document Management	Executive Secretaries and Administrative Assistants	http://www.bls.gov/oes/2002/naics4_541300.htm	17.67	35.34
Eng	Enterprise Resource Planning	Management Analysts	http://www.bls.gov/oes/2002/naics4_541300.htm	35.11	70.22
Eng	Facility Planning and Scheduling	Civil Engineers	http://www.bls.gov/oes/2002/naics4_541300.htm	30.53	61.06
Eng	Facility Simulation	Civil Engineers	http://www.bls.gov/oes/2002/naics4_541300.htm	30.53	61.06
Eng	Information Requests	Executive Secretaries and Administrative Assistants	http://www.bls.gov/oes/2002/naics4_541300.htm	17.67	35.34
Eng	Materials Management	Production, Planning, and Expediting Clerks	http://www.bls.gov/oes/2002/naics4_541300.htm	18.73	37.46
Eng	Procurement	Purchasing Agents, Except Wholesale, Retail, and Farm Products	http://www.bls.gov/oes/2002/naics4_541300.htm	25.78	51.56
Eng	Project Management	Civil Engineers	http://www.bls.gov/oes/2002/naics4_541300.htm	30.53	61.06
Eng	Accountant	Accountants and Auditors	http://www.bls.gov/oes/2002/naics4_541300.htm	25.49	50.98

Table C-2. Wage Rates for General Contractors

Labor Category	BLS Title	Source	Wage (2002)	Loaded (2002)
CAx User	Civil Engineers	http://www.bls.gov/oes/oes/2002/naics4_236200.htm	28.57	57.14
Design Support Specialist	Civil Engineers	http://www.bls.gov/oes/oes/2002/naics4_236200.htm	28.57	57.14
Software Support Specialist	Computer Support Specialists	http://www.bls.gov/oes/oes/2002/naics4_236200.htm	24.52	49.04
Network and Systems Administrators	Network and Computer Systems Administrators	http://www.bls.gov/oes/oes/2002/naics4_236200.htm	26.32	52.64
Cost Estimation	Cost Estimators	http://www.bls.gov/oes/oes/2002/naics4_236200.htm	27.78	55.56
Document Management	Executive Secretaries and Administrative Assistants	http://www.bls.gov/oes/oes/2002/naics4_236200.htm	16.83	33.66
Enterprise Resource Planning	Management Analysts	http://www.bls.gov/oes/oes/2002/naics4_236200.htm	31.32	62.64
Facility Planning and Scheduling	Civil Engineers	http://www.bls.gov/oes/oes/2002/naics4_236200.htm	28.57	57.14
Facility Simulation	Civil Engineers	http://www.bls.gov/oes/oes/2002/naics4_236200.htm	28.57	57.14
Information Requests	Executive Secretaries and Administrative Assistants	http://www.bls.gov/oes/oes/2002/naics4_236200.htm	16.83	33.66
Inspection & Certification	Civil Engineers	http://www.bls.gov/oes/oes/2002/naics4_236200.htm	28.57	57.14
Maintenance Planning and Management	Civil Engineers	http://www.bls.gov/oes/oes/2002/naics4_236200.htm	28.57	57.14
Materials Management	Production, Planning, and Expediting Clerks	http://www.bls.gov/oes/oes/2002/naics4_236200.htm	18.13	36.26
Procurement	Purchasing Agents, Except Wholesale, Retail, and Farm Products	http://www.bls.gov/oes/oes/2002/naics4_236200.htm	25.29	50.58
Product Data Management	Management Analysts	http://www.bls.gov/oes/oes/2002/naics4_236200.htm	31.32	62.64
Project Management	Construction Manager	http://www.bls.gov/oes/oes/2002/naics4_236200.htm	34.33	68.66
Worker	Construction Laborer	http://www.bls.gov/oes/oes/2002/naics4_236200.htm	14.72	29.44
Startup and Commissioning	Civil Engineers	http://www.bls.gov/oes/oes/2002/naics4_236200.htm	28.57	57.14
Accounting	Accountants and Auditors	http://www.bls.gov/oes/oes/2002/naics4_236200.htm	25.75	51.5

Table C-3. Wage Rates for Specialty Fabricators and Suppliers

Labor Category	BLS Title	Source	Wage (2002)	Loaded (2002)
CAx User	Civil Engineers	http://www.bls.gov/oes/2002/naics4_238200.htm	27.20	54.40
Design Support Specialist	Civil Engineers	http://www.bls.gov/oes/2002/naics4_238200.htm	27.20	54.40
Software Support Specialist	Computer Support Specialists	http://www.bls.gov/oes/2002/naics4_238200.htm	22.05	44.10
Network and Systems Administrators	Network and Computer Systems Administrators	http://www.bls.gov/oes/2002/naics4_238200.htm	27.08	54.16
Cost Estimation	Cost Estimators	http://www.bls.gov/oes/2002/naics4_238200.htm	25.61	51.22
Document Management	Executive Secretaries and Administrative Assistants	http://www.bls.gov/oes/2002/naics4_238200.htm	15.89	31.78
Enterprise Resource Planning	Management Analysts	http://www.bls.gov/oes/2002/naics4_238200.htm	27.81	55.62
Facility Planning and Scheduling	Civil Engineers	http://www.bls.gov/oes/2002/naics4_238200.htm	27.20	54.40
Facility Simulation	Civil Engineers	http://www.bls.gov/oes/2002/naics4_238200.htm	27.20	54.40
Information Requests	Executive Secretaries and Administrative Assistants	http://www.bls.gov/oes/2002/naics4_238200.htm	15.89	31.78
Inspection & Certification	Civil Engineers	http://www.bls.gov/oes/2002/naics4_238200.htm	27.20	54.40
Maintenance Planning and Management	Civil Engineers	http://www.bls.gov/oes/2002/naics4_238200.htm	27.20	54.40
Materials Management	Production, Planning, and Expediting Clerks	http://www.bls.gov/oes/2002/naics4_238200.htm	17.21	34.42
Procurement	Purchasing Agents, Except Wholesale, Retail, and Farm Products	http://www.bls.gov/oes/2002/naics4_238200.htm	21.33	42.66
Product Data Management	Management Analysts	http://www.bls.gov/oes/2002/naics4_238200.htm	27.81	55.62
Project Management	Construction Manager	http://www.bls.gov/oes/2002/naics4_238200.htm	36.37	72.74
Worker	Construction Laborer	http://www.bls.gov/oes/2002/naics4_238200.htm	14.01	28.02
Startup and Commissioning	Civil Engineers	http://www.bls.gov/oes/2002/naics4_238200.htm	27.20	54.40
Accounting	Accountants and Auditors	http://www.bls.gov/oes/2002/naics4_238200.htm	25.45	50.90

Table C-4. Wage Rates for Owners and Operators

Labor Category	BLS Title	Source	Wage (2002)	Loaded (2002)
CAx User	Architects, Except Landscape and Naval	http://www.bls.gov/oes/oes/2002/oes_nat.htm	30.06	60.12
Design Support Specialist	Architects, Except Landscape and Naval	http://www.bls.gov/oes/oes/2002/oes_nat.htm	30.06	60.12
Software Support Specialist	Computer Support Specialists	http://www.bls.gov/oes/oes/2002/oes_nat.htm	20.35	40.70
Network and Systems Administrators	Network and Computer Systems Administrators	http://www.bls.gov/oes/oes/2002/oes_nat.htm	27.70	55.40
Cost Estimation	Cost Estimators	http://www.bls.gov/oes/oes/2002/oes_nat.htm	24.67	49.34
Document Management	Executive Secretaries and Administrative Assistants	http://www.bls.gov/oes/oes/2002/oes_nat.htm	16.85	33.70
Enterprise Resource Planning	Management Analysts	http://www.bls.gov/oes/oes/2002/oes_nat.htm	33.73	67.46
Facility Planning and Scheduling	Civil Engineers	http://www.bls.gov/oes/oes/2002/oes_nat.htm	30.29	60.58
Facility Simulation	Civil Engineers	http://www.bls.gov/oes/oes/2002/oes_nat.htm	30.29	60.58
Inspection and Certification	Civil Engineers	http://www.bls.gov/oes/oes/2002/oes_nat.htm	30.29	60.58
Information Requests	Executive Secretaries and Administrative Assistants	http://www.bls.gov/oes/oes/2002/oes_nat.htm	16.85	33.70
Maintenance Planning & Management	Civil Engineers	http://www.bls.gov/oes/oes/2002/oes_nat.htm	30.29	60.58
Materials Management	Production, Planning, and Expediting Clerks	http://www.bls.gov/oes/oes/2002/oes_nat.htm	16.87	33.74
Procurement	Purchasing Agents, Except Wholesale, Retail, and Farm Products	http://www.bls.gov/oes/oes/2002/oes_nat.htm	23.21	46.42
Product Data Management	Management Analysts	http://www.bls.gov/oes/oes/2002/oes_nat.htm	33.73	67.46
Project Management	Architects, Except Landscape and Naval	http://www.bls.gov/oes/oes/2002/oes_nat.htm	30.06	60.12
Start-up & Commissioning	Civil Engineers	http://www.bls.gov/oes/oes/2002/oes_nat.htm	30.29	60.58
Accounting	Accounting and Auditors	http://www.bls.gov/oes/oes/2002/oes_nat.htm	25.59	51.18
Facilities Management Staffer	Civil Engineering Technicians	http://www.bls.gov/oes/oes/2002/oes_nat.htm	18.71	37.42

Appendix D:
Inadequate Interoperability Cost Variability for Owners and Operators

Table D-1. Inadequate Interoperability Cost Variability for Owners and Operators: Key Cost Components

Life-Cycle Phase	Cost Category	Cost Component	Average Cost per Square Foot	Average Cost per Square Meter	25th Percentile (Square Foot)	25th Percentile (Square Meter)	50th Percentile (Square Foot)	50th Percentile (Square Meter)	75th Percentile (Square Foot)	75th Percentile (Square Meter)	Standard Deviation	Standard Deviation
Planning, Engineering, and Design Phase	Avoidance Costs	Inefficient Business Process Managament Costs	0.37826	4.07155	0.09493	1.02178	0.14141	1.52212	0.21227	2.28480	0.75841	8.16344
		Interoperability Research and Development Expenditures	0.00389	0.04186	0.00000	0.00000	0.00147	0.01582	0.00378	0.04069	0.00300	0.03228
	Mitigation Costs	Manual Reentry Costs	0.15556	1.67442	0.13400	1.44238	0.18036	1.94138	0.30819	3.31736	0.17519	1.88569
		Design and Construction Information Verification Costs	0.00564	0.06072	0.00077	0.00828	0.01156	0.12442	0.05551	0.59751	0.07107	0.76494
		RFI Management Costs	0.09231	0.99364	0.01052	0.11325	0.01924	0.20708	0.15872	1.70841	0.16419	1.76737
	Subtotal	Avoidance Costs	0.38215	4.07155	0.09493	1.02178	0.14435	1.55377	0.21856	2.35260	0.75761	8.15487
		Mitigation Costs	0.25351	2.72877	0.30854	3.32106	0.36072	3.88276	0.44621	4.80297	0.17524	1.88628
		Subtotal	0.63566	6.80032	0.39256	4.22548	0.51669	5.56155	0.62429	6.71977	0.68787	7.40418
Construction Phase	Avoidance Costs	Inefficient Business Process Managament Costs	0.49418	5.31934	0.00974	0.10483	0.07856	0.84562	0.17367	1.86939	1.44536	15.55769
		Interoperability Research and Development Expenditures	0.00318	0.03425	0.00000	0.00000	0.00120	0.01295	0.00309	0.03329	0.00245	0.02641
	Mitigation Costs	Manual Reentry Costs	0.14772	1.59009	0.01899	0.20436	0.13467	1.44956	0.22061	2.37467	0.21398	2.30323
		Design and Construction Information Verification Costs	0.00677	0.07290	0.00000	0.00000	0.00000	0.00000	0.02255	0.24267	0.02777	0.29894
		RFI Management Costs	0.13789	1.48425	0.00069	0.00741	0.04810	0.51770	0.15872	1.70841	0.13978	1.50460
	Subtotal	Avoidance Costs	0.49736	5.31934	0.06316	0.67985	0.12990	1.39819	0.95754	10.30686	1.60342	17.25904
		Mitigation Costs	0.29239	3.14723	0.30138	3.24401	0.36382	3.91611	0.56467	6.07809	0.22184	2.38783
		Subtotal	0.78975	8.46658	0.39029	4.20109	0.55038	5.92421	0.69425	7.47288	1.33070	14.32350

(continued)

Table D-1. Inadequate Interoperability Cost Variability for Owners and Operators: Key Cost Components (Continued)

Life-Cycle Phase	Cost Category	Cost Component	Average Cost per Square Foot	Average Cost per Square Meter	25th Percentile (Square Foot)	25th Percentile (Square Meter)	50th Percentile (Square Foot)	50th Percentile (Square Meter)	75th Percentile (Square Foot)	75th Percentile (Square Meter)	Standard Deviation (Square Foot)	Standard Deviation (Square Meter)
Operations and Maintenance Phase	Avoidance Costs	Inefficient Business Process Management Costs	0.04240	0.45644	0.00027	0.00287	0.00622	0.06697	0.31044	3.34156	0.60070	6.46590
		Productivity Loss and Training Costs on Redundant Facility Management Systems	0.00033	0.00351	0.00000	0.00000	0.00000	0.00000	0.00000	0.00000	0.00072	0.00775
		Redundant Facility Management Systems IT Support Staffing Costs	0.00028	0.00298	0.00000	0.00000	0.00000	0.00000	0.00115	0.01234	0.00229	0.02469
		Interoperability Research and Development Expenditures	0.00004	0.00046	0.00000	0.00000	0.00001	0.00011	0.00010	0.00113	0.00018	0.00190
	Mitigation Costs	O&M Staff Productivity Loss	0.01587	0.17081	0.00805	0.08669	0.01611	0.17338	0.01928	0.20752	0.01157	0.12458
		O&M Staff Rework Costs	0.00010	0.00110	0.00053	0.00575	0.00107	0.01151	0.00160	0.01726	0.00151	0.01627
		O&M Information Verification Costs	0.12394	1.33406	0.04510	0.48549	0.12873	1.38568	0.22584	2.43096	0.13235	1.42462
	Delay Costs	Idled Employees Costs	0.03881	0.41771	0.01871	0.20139	0.03742	0.40279	0.03884	0.41811	0.02247	0.24188
	Subtotal	Avoidance Costs	0.05485	0.58995	0.00943	0.10149	0.01442	0.15518	0.31367	3.37626	0.59792	6.43596
		Mitigation Costs	0.13991	1.50597	0.00000	0.00000	0.01202	0.12943	0.08072	0.86887	0.09525	1.02530
		Delay Costs	0.03881	0.41771	0.01871	0.20139	0.03742	0.40279	0.03884	0.41811	0.02247	0.24188
		Subtotal	0.23357	2.51362	0.02140	0.23037	0.15182	1.63414	0.53733	5.78373	0.62350	6.71133

Note: The variability of costs for redundant facilities management systems is not presented in Appendix D to prevent the disclosure of individual survey responses.

Source: RTI Estimates.

www.ingramcontent.com/pod-product-compliance
Lightning Source LLC
Chambersburg PA
CBHW081442170526
45166CB00008B/2284